Development and Application of Light-Field Cameras in Fluid Measurements

Shengxian Shi · T. H. New
Editors

Development and Application of Light-Field Cameras in Fluid Measurements

Springer

Editors
Shengxian Shi
School of Mechanical Engineering
Shanghai Jiao Tong University
Shanghai, China

T. H. New
School of Mechanical and Aerospace Engineering
Nanyang Technological University
Singapore, Singapore

ISBN 978-3-031-01778-0   ISBN 978-3-031-01779-7 (eBook)
https://doi.org/10.1007/978-3-031-01779-7

This Springer imprint is published by the registered company Springer Nature Switzerland AG
The registered company address is: Gewerbestrasse 11, 6330 Cham, Switzerland

Preface

Humans have evolved to possess acute pseudo-3D perception via stereoscopic imaging of our eyesight, and it should not come as a surprise that we desire to capture 3D information in our scientific research and applications as rapidly, conveniently and cost-effectively as well. However, unlike our brains which have evolved to allow us to deduce and infer 3D information swiftly, achieving 3D visualisations and measurements in engineering research and applications is much more challenging. A multitude of technical challenges needs to be solved, in terms of optical principles, imaging technology, advanced 3D post-processing algorithms and computing prowess, among others. They become particularly more severe if a 3D imaging and measurement system meant for flow measurements were to be compact, user-friendly and priced competitively relative to its capabilities. System costs tend to be a significant barrier to entry for many research institutions, universities and corporations, which prevent more rapid and widespread adoption. This in turn could have an undesirable outcome of impeding more rapid progress in our collective research advances.

Earlier efforts primarily focused on utilising the principles of stereoscopic human eyesight where two imaging devices capture the same scene from different perspectives. With accurate knowledge of the optical light paths and principles, system configurations, hardware capabilities and proper calibrations, information in the depth direction may be obtained to reconstruct a pseudo-3D understanding. For flow measurements, this can be seen in stereoscopic particle image velocimetry systems that were introduced commercially more than two decades ago, which continue to be relevant today. However, as more complex flow scenarios and configurations evolved over the years, more truly 3D flow measurements were needed. This is especially the case for scenarios that are either highly 3D, offer limited optical access for imaging or turbulent combustion related, among many other challenging ones. To overcome some of these technical limitations and/or requirements, three- and four-camera-based volumetric particle image velocimetry principles and systems were researched upon and the latter four-camera-based volumetric particle image velocimetry (more commonly known as tomographic particle image velocimetry) configuration become one of the most popular 3D flow measurement approaches today. Its working principles are now well established and shown to be highly adaptable to many challenging flow measurement scenarios. On the other hand, the use of a four-camera configuration with each of them imaging the flow scenario from different perspectives, introduces additional obstacles as well. Despite rapid advancements in camera technology at lower price points, a tomographic particle image velocimetry remains relatively expensive with the use of four cameras and all the accompanying accessories. Hence, it is at this point not as cost-effective and accessible as more conventional 2D or stereoscopic particle image velocimetry approaches, certainly not for laboratories with moderate resources. Additionally, the complexities associated with setting up tomographic particle image velocimetry systems are non-trivial and require skilled experimentalists. This is on top of the significant time required to post-process images taken from the four cameras for reconstruction of 3D volumetric flow information, though this can be circumvented by the use of high-performance computing facilities. However, this may not be readily accessible to all tomographic particle image velocimetry operators.

Clearly, new approaches toward 3D imaging-based flow measurements are needed, if significant improvements to the cost, complexity and post-processing time were to be made. Indeed, many researchers are now beginning to explore the feasibility of utilising light-field cameras for such measurements, where it may be possible to evolve the four-camera-based tomographic particle image velocimetry approach into one that requires only one light-field camera. A light-field camera essentially comprises of an additional and typically hemispherical microlens array layer located in front of the traditional 2D imaging sensor, and studies have shown that it is possible to extract 3D information through the implementation of "Light-field" principles. With the need for only one light-field camera instead of multiple cameras, challenging 3D flow measurements could be made at significantly lower cost and complexity levels, which could drive higher adoption rates in the research community. Therefore, it should not come as a surprise then that research and implementation of light-field-based particle image velocimetry approaches are gaining momentum as we speak, and that the present book will be a timely contribution toward understanding the fundamental working principles, light-field camera hardware, calibration procedures, light-field particle image velocimetry principles and assessments on the flow measurement accuracy levels. On top of that, case studies on the practical implementation of light-field particle image velocimetry across a wide range of flow scenarios, the use of light-field cameras in surface pressure measurements and a brief coverage of how light-field cameras may prove to be useful for other research and engineering applications will also be covered.

Last but not least, the editors would like to extend their deep appreciation to all authors who had contributed toward their respective chapters. Without their contributions, support and enthusiasm throughout the preparation stage, this book would not have been possible.

Shanghai, China
Singapore
January 2022

Shengxian Shi
T. H. New

Contents

Contributors

Ganapathisubramani B. Aerodynamics and Flight Mechanics Research Group, University of Southampton, Southampton, UK

New T. H. School of Mechanical and Aerospace Engineering, Nanyang Technological University, Singapore, Singapore

Puayen Tan Zu Department of Mechanical Engineering, National Yang Ming Chiao Tung University, Hsinchu, Taiwan

Quinn Mark Kenneth Department of Mechanical, Aerospace and Civil Engineering, University of Manchester, Manchester, UK

Shi Shengxian School of Mechanical Engineering, Shanghai Jiao Tong University, Shanghai, China

Soria J. Laboratory for Turbulence Research in Aerospace and Combustion, Department of Mechanical and Aerospace Engineering, Monash University, Melbourne, VIC, Australia

Thurow Brian Department of Aerospace Engineering, Auburn University, Auburn, AL, USA

Zhao Zhou School of Mechanical Engineering, Shanghai Jiao Tong University, Shanghai, China

Abbreviations

1D	One-Dimensional
2D	Two-Dimensional
2D-3C	Two-Dimensional Three-Component
2D-PIV	Two-Dimensional PIV
3D	Three-Dimensional
3D-3C	Three-Dimensional Three-Component
3D-PIV	Three-Dimensional PIV, Or Volumetric PIV
4D	Four-Dimensional
AOA	Angles-Of-Attack
APG	Adverse Pressure Gradient
AR	Aspect Ratio
ART	Algebraic Reconstruction Technique
BOS	Background-Oriented Schlieren
CCD	Charge Coupled Device
CDF	Cumulative Distribution Function
CFD	Computational Fluid Dynamics
CMM	Coordinate-Measuring Machines
CMOS	Complementary Metal Oxide Semiconductor
CNC	Computerised Numerical Control
CoC	Circle of Confusion
CTC	Computed Tomography of Chemiluminescence
CUDA	Compute Unified Device Architecture
DDPIV	Defocusing Digital PIV
DMD	Dynamic Mode Decomposition
DNS	Direct Numerical Simulations
DOF	Depth-Of-Field
DRT-MART	Dense Ray Tracing Multiplicative Algebraic Reconstruction Technique
DSLR	Digital Single-Lens Reflex
EoC	Ellipse of Confusion
EPI	Epipolar-Plane-Image

FFL	Flange Focal Length
FPS	Frame Per Second
FSI	Fluid-Structure-Interaction
GP	Gradient Projection
GPU	Graphics Processing Unit
HPIV	Holographic PIV
HSST	Intermittent Blow-Down High Supersonic Tunnel
IR	Infrared Radiation
JPDF	Joint Probability Distribution Functions
kHz	Kilohertz
LCS	Lens Coordinate System
LED	Light Emitting Diode
LF	Light-Field
LF-MRT	LF Multispectral Radiation Thermometry
LF-PIV	Light-Field Particle Image Velocimetry
LSV	Laser Speckle Velocimetry
LTPR	Light-field Camera To Tomo-Camera Pixel Ratio
LTRAC	Laboratory for Turbulence Research In Aerospace And Combustion
MART	Multiplicative Reconstruction Technique
MLA	Microlens Array
MLOS	Multiplicative Line-Of-Sight
MRT	Multi-Spectral Radiation Thermometry
NPR	Nozzle-Pressure Ratio
PCS	2D Pixel Coordinate System
PDF	Probability Distribution Functions
PIV	Particle Image Velocimetry
PMR	Pixel-to-Microlens Ratio
POD	Proper Orthogonal Decomposition
PPM	Particle Per Microlens
PPP	Particle Per Pixel
PSF	Point Source Response
PSP	Pressure-Sensitive Paint
PTF	Parallel-To-Film
PVR	Pixel Voxel Ratio
RGB	Red Green Blue
RMS	Root Mean Square
RTM	Ray Transfer Matrix
SAPIV	Synthetic Aperture PIV
SLI	Scalable Link Interface
SQP	Sequence Quadratic Program
Stereo-PIV	Stereoscopic PIV
SV-I and SV-II	Streamwise Vortex I and II
TBL	Turbulent Boundary Layer
TKE	Turbulent Kinetic Energy

Tomo-PIV	Tomographic PIV
TR-PIV	Time-Resolved Particle Image Velocimetry
UV	Ultraviolet
VR	Virtualized Reality

Symbols

I	Pixel intensity
x, y, z	Spatial coordinate
θ, φ	Angular coordinate
λ	Wavelength
$w_{i,j}$	Weighting coefficient for MART reconstruction
C_{df}	Center of the plenoptic disc feature (in pixel plane)
D_{df}	Diameter of the plenoptic disc feature (in pixel plane)
p_c	Center of point-like feature
f_m	Focal length of the main lens
$f_{\#}, f/\#$	f-number
p_m	Effective size of the main lens
S_o	Distance from the object focal plane to the main lens
S_i	Distance from the main lens to the image focal plane
M	Magnification of the plenoptic camera, $M = -S_o/S_i$
γ	Tilted-angle, the angle between main lens plane and MLA plane
α	The angle between main lens plane and focal plane, $\tan\alpha = -M\tan\gamma$
f_l	Focal length of the lenslet
p_l	Lenslet pitch
p_p	Pixel pitch
O	Ideal optical center (of the main lens)
O_m	Intersection point between optical axis and MLA plane
O_p	Perpendicular projected point of O_m on sensor plane
A	Aperture center
A_1	Perpendicular projected point of A on MLA plane
A_m	Size of aperture
P	Object point, a point light source
Q	Converging image point
Q_1	Perpendicular projected point of Q on MLA plane

R	Intersection point of the MLA plane and the line that passes through A and Q
p_{ci}	Center of the i-th point-like pattern beneath i-th lenslet
h_{ci}	Center of the i-th lenslet on the senor plane
$C_{l(i)}$	Coordinate of the i-th lenslet center in pixel plane
g_{ci}	Center of the i-th aperture projection point beneath i-th lenslet
l_{ci}	Center of the i-th lenslet on MLA plane
p_c	Center of the point-like pattern beneath a lenslet
h_c	Center of a lenslet on the senor plane
g_c	Center of the aperture projection point beneath a lenslet
l_c	Center of a lenslet on MLA plane
C_{EoC}	Center of the EoC formed on MLA plane
D_{EoC}^{major}	Length of the EoC's major axis
D_{EoC}^{minor}	Length of the EoC's minor axis
D_{CoC}	Dimeter of the CoC
$(x, y, z)_o$	Coordinates of an arbitrary point source in the object space
$(x, y, z)_i$	Coordinates of the image of an arbitrary point source $(x, y, z)_o$ in the image space
D_a	Diameter of the main lens
$z_{o,far}$	DOF limit on the far-side of the focal-plane
$z_{o,near}$	DOF limit on the near-side of the focal-plane
$L(s, t, u, v)$	The 4D light-field, where (s, t) and (u, t) respectively define the intersection location of a ray made with two parallel planes
$L(x, y, \theta, \varphi)$	The 4D light-field, where (x, y) and (θ, φ) respectively define the spatial- and angular-coordinates of a ray
f_l	Focal length of the microlens
$d_{i,l}$	Distance from the MLA plane to the sensor plane
s'	A virtual image point
$d_{o,l}$	Distance from the image-plane to the MLA plane
$E(X_j, Y_j, Z_j)$	Intensity of a voxel at (X_j, Y_j, Z_j)
$I(x_i, y_i)$	Pixel intensity of a light-field image
(X_j, Y_j, Z_j)	Coordinates of the j-th voxel
$(x_i, y_i)(i \in N_j)$	Coordinates of the i-th pixel, N_j is the number of object point influential on the pixel
$w_{i,j}$	Weighting coefficient for MART reconstruction
C_{df}	Center of the plenoptic disc feature (in pixel plane)
D_{df}	Diameter of the plenoptic disc feature (in pixel plane)
y_c	Intersection of a light ray with the main lens, light ray passes lenslet center
p_{l_sensor}	Lenslet pitch (in sensor plane), $p_{l\_sensor} = p_l \frac{S_i+f_l}{S_i} = \left\|C_{l(i)} - C_{l(i-1)}\right\|$

$\mathcal{M}_{4\times 3}$	Mapping matrix that relates a spatial point $P(X, Y, Z)$ With its corresponding C_{df}
w_1	Lenslet weighting coefficient
w_2	Pixel weighting coefficient
y_l	Offset of a light ray from the lenslet center
y_m	Intersection of a light ray made with the main lens
y_p	Intersection of a light ray made with the pixel plane
$O - XYZ$	Lens coordinate system (LCS)
$O_p - UV$	2D pixel coordinate system (PCS)
$S_i^{'}$	The Z-direction distance from Q to main lens
Q_d	Distance between Q and Q_1
$\begin{pmatrix} \mu_{up} \\ \sigma_{up} \end{pmatrix}$	The angle that the light ray $\underset{QI^{up}}{\rightarrow}$ made with MLA plane
γ_x	Sensor rotation angle along X-axis
γ_y	Sensor rotation angle along Y-axis
A^d	Distance AA_1
k	Correction factor of the major/minor axis length of EoC
a_{0-19}	Polynomial coefficients
L_s	Scheimpflug line
Q_{Recon}	Reconstruction quality
$E_0(x, y, z)$	Voxel intensity approximated by a Gaussian distribution with three voxel diameters
$E_1(x, y, z)$	Voxel intensity of the reconstructed particle image
μ	Relaxation factor
ω	Angular frequency
Re	Reynolds number
I_{ref}	Pixel intensity of images captured at wind-off condition
I	The pixel intensity of images captured at wind-on condition
$A(t), B(t)$	Paint calibration coefficients which are normally dependent on local temperature
P_{windon}	Model surface pressure distribution at wind-on condition
P_{ref}	The model surface pressure distribution at wind-off condition
dP/dx	The streamwise pressure gradient
δ_1	The boundary layer displacement thickness
τ_w	The local mean wall shear stress
$(u, v, w), (U, V, W)$	Velocity in three directions
f	Frequency
$u_{\max}$	Maximum velocity encountered during the blowing cycle
τ	Period of the cycle
$\bar{u}$	Mean blowing velocity
U_0	Momentum flow velocity
d	Orifice geometry width
l	Orifice geometry length
Sr	Strouhal number

L_0	Stroke length
D_{so}	Saddle location
$\boldsymbol{D}$	Velocity gradient tensor
λ_{ci}	Swirl strength
c	Blade chord length
c_x	Blade axial chord length
p	Blade pitch
h	Blade span
β_1	Inlet design flow angle
β_2	Outlet design flow angle
Ma	Mach number
Re_c	Reynolds number based on the blade chord length, c
κ_1, κ_2	Principal curvatures
κ_m, κ_G	Mean and Gaussian curvatures
I_i	Radiation intensity of i-th channel
λ_i	Effective wavelength of the i-th channel
A_i	Calibration factor which depends on the wavelength λ_i
T	True Temperature in Eq. (7.1)
C_2	The second radiation constant
$\varepsilon(\lambda_i, T)$	The spectral emissivity of the target at true temperature T
(S_x, S_y)	Lenslet center (relative to optical center)
C_{pk}	Process capability index

Development and Application of Light-Field Cameras in Fluid Measurements

Shengxian Shi · T. H. New
Editors

Development and Application of Light-Field Cameras in Fluid Measurements

Springer

Editors
Shengxian Shi
School of Mechanical Engineering
Shanghai Jiao Tong University
Shanghai, China

T. H. New
School of Mechanical and Aerospace Engineering
Nanyang Technological University
Singapore, Singapore

ISBN 978-3-031-01778-0 ISBN 978-3-031-01779-7 (eBook)
https://doi.org/10.1007/978-3-031-01779-7

This Springer imprint is published by the registered company Springer Nature Switzerland AG
The registered company address is: Gewerbestrasse 11, 6330 Cham, Switzerland

Preface

Humans have evolved to possess acute pseudo-3D perception via stereoscopic imaging of our eyesight, and it should not come as a surprise that we desire to capture 3D information in our scientific research and applications as rapidly, conveniently and cost-effectively as well. However, unlike our brains which have evolved to allow us to deduce and infer 3D information swiftly, achieving 3D visualisations and measurements in engineering research and applications is much more challenging. A multitude of technical challenges needs to be solved, in terms of optical principles, imaging technology, advanced 3D post-processing algorithms and computing prowess, among others. They become particularly more severe if a 3D imaging and measurement system meant for flow measurements were to be compact, user-friendly and priced competitively relative to its capabilities. System costs tend to be a significant barrier to entry for many research institutions, universities and corporations, which prevent more rapid and widespread adoption. This in turn could have an undesirable outcome of impeding more rapid progress in our collective research advances.

Earlier efforts primarily focused on utilising the principles of stereoscopic human eyesight where two imaging devices capture the same scene from different perspectives. With accurate knowledge of the optical light paths and principles, system configurations, hardware capabilities and proper calibrations, information in the depth direction may be obtained to reconstruct a pseudo-3D understanding. For flow measurements, this can be seen in stereoscopic particle image velocimetry systems that were introduced commercially more than two decades ago, which continue to be relevant today. However, as more complex flow scenarios and configurations evolved over the years, more truly 3D flow measurements were needed. This is especially the case for scenarios that are either highly 3D, offer limited optical access for imaging or turbulent combustion related, among many other challenging ones. To overcome some of these technical limitations and/or requirements, three- and four-camera-based volumetric particle image velocimetry principles and systems were researched upon and the latter four-camera-based volumetric particle image velocimetry (more commonly known as tomographic particle image velocimetry) configuration become one of the most popular 3D flow measurement approaches today. Its working principles are now well established and shown to be highly adaptable to many challenging flow measurement scenarios. On the other hand, the use of a four-camera configuration with each of them imaging the flow scenario from different perspectives, introduces additional obstacles as well. Despite rapid advancements in camera technology at lower price points, a tomographic particle image velocimetry remains relatively expensive with the use of four cameras and all the accompanying accessories. Hence, it is at this point not as cost-effective and accessible as more conventional 2D or stereoscopic particle image velocimetry approaches, certainly not for laboratories with moderate resources. Additionally, the complexities associated with setting up tomographic particle image velocimetry systems are non-trivial and require skilled experimentalists. This is on top of the significant time required to post-process images taken from the four cameras for reconstruction of 3D volumetric flow information, though this can be circumvented by the use of high-performance computing facilities. However, this may not be readily accessible to all tomographic particle image velocimetry operators.

Clearly, new approaches toward 3D imaging-based flow measurements are needed, if significant improvements to the cost, complexity and post-processing time were to be made. Indeed, many researchers are now beginning to explore the feasibility of utilising light-field cameras for such measurements, where it may be possible to evolve the four-camera-based tomographic particle image velocimetry approach into one that requires only one light-field camera. A light-field camera essentially comprises of an additional and typically hemispherical microlens array layer located in front of the traditional 2D imaging sensor, and studies have shown that it is possible to extract 3D information through the implementation of "Light-field" principles. With the need for only one light-field camera instead of multiple cameras, challenging 3D flow measurements could be made at significantly lower cost and complexity levels, which could drive higher adoption rates in the research community. Therefore, it should not come as a surprise then that research and implementation of light-field-based particle image velocimetry approaches are gaining momentum as we speak, and that the present book will be a timely contribution toward understanding the fundamental working principles, light-field camera hardware, calibration procedures, light-field particle image velocimetry principles and assessments on the flow measurement accuracy levels. On top of that, case studies on the practical implementation of light-field particle image velocimetry across a wide range of flow scenarios, the use of light-field cameras in surface pressure measurements and a brief coverage of how light-field cameras may prove to be useful for other research and engineering applications will also be covered.

Last but not least, the editors would like to extend their deep appreciation to all authors who had contributed toward their respective chapters. Without their contributions, support and enthusiasm throughout the preparation stage, this book would not have been possible.

Shanghai, China
Singapore
January 2022

Shengxian Shi
T. H. New

Contents

Contributors

Ganapathisubramani B. Aerodynamics and Flight Mechanics Research Group, University of Southampton, Southampton, UK

New T. H. School of Mechanical and Aerospace Engineering, Nanyang Technological University, Singapore, Singapore

Puayen Tan Zu Department of Mechanical Engineering, National Yang Ming Chiao Tung University, Hsinchu, Taiwan

Quinn Mark Kenneth Department of Mechanical, Aerospace and Civil Engineering, University of Manchester, Manchester, UK

Shi Shengxian School of Mechanical Engineering, Shanghai Jiao Tong University, Shanghai, China

Soria J. Laboratory for Turbulence Research in Aerospace and Combustion, Department of Mechanical and Aerospace Engineering, Monash University, Melbourne, VIC, Australia

Thurow Brian Department of Aerospace Engineering, Auburn University, Auburn, AL, USA

Zhao Zhou School of Mechanical Engineering, Shanghai Jiao Tong University, Shanghai, China

Abbreviations

1D	One-Dimensional
2D	Two-Dimensional
2D-3C	Two-Dimensional Three-Component
2D-PIV	Two-Dimensional PIV
3D	Three-Dimensional
3D-3C	Three-Dimensional Three-Component
3D-PIV	Three-Dimensional PIV, Or Volumetric PIV
4D	Four-Dimensional
AOA	Angles-Of-Attack
APG	Adverse Pressure Gradient
AR	Aspect Ratio
ART	Algebraic Reconstruction Technique
BOS	Background-Oriented Schlieren
CCD	Charge Coupled Device
CDF	Cumulative Distribution Function
CFD	Computational Fluid Dynamics
CMM	Coordinate-Measuring Machines
CMOS	Complementary Metal Oxide Semiconductor
CNC	Computerised Numerical Control
CoC	Circle of Confusion
CTC	Computed Tomography of Chemiluminescence
CUDA	Compute Unified Device Architecture
DDPIV	Defocusing Digital PIV
DMD	Dynamic Mode Decomposition
DNS	Direct Numerical Simulations
DOF	Depth-Of-Field
DRT-MART	Dense Ray Tracing Multiplicative Algebraic Reconstruction Technique
DSLR	Digital Single-Lens Reflex
EoC	Ellipse of Confusion
EPI	Epipolar-Plane-Image

FFL	Flange Focal Length
FPS	Frame Per Second
FSI	Fluid-Structure-Interaction
GP	Gradient Projection
GPU	Graphics Processing Unit
HPIV	Holographic PIV
HSST	Intermittent Blow-Down High Supersonic Tunnel
IR	Infrared Radiation
JPDF	Joint Probability Distribution Functions
kHz	Kilohertz
LCS	Lens Coordinate System
LED	Light Emitting Diode
LF	Light-Field
LF-MRT	LF Multispectral Radiation Thermometry
LF-PIV	Light-Field Particle Image Velocimetry
LSV	Laser Speckle Velocimetry
LTPR	Light-field Camera To Tomo-Camera Pixel Ratio
LTRAC	Laboratory for Turbulence Research In Aerospace And Combustion
MART	Multiplicative Reconstruction Technique
MLA	Microlens Array
MLOS	Multiplicative Line-Of-Sight
MRT	Multi-Spectral Radiation Thermometry
NPR	Nozzle-Pressure Ratio
PCS	2D Pixel Coordinate System
PDF	Probability Distribution Functions
PIV	Particle Image Velocimetry
PMR	Pixel-to-Microlens Ratio
POD	Proper Orthogonal Decomposition
PPM	Particle Per Microlens
PPP	Particle Per Pixel
PSF	Point Source Response
PSP	Pressure-Sensitive Paint
PTF	Parallel-To-Film
PVR	Pixel Voxel Ratio
RGB	Red Green Blue
RMS	Root Mean Square
RTM	Ray Transfer Matrix
SAPIV	Synthetic Aperture PIV
SLI	Scalable Link Interface
SQP	Sequence Quadratic Program
Stereo-PIV	Stereoscopic PIV
SV-I and SV-II	Streamwise Vortex I and II
TBL	Turbulent Boundary Layer
TKE	Turbulent Kinetic Energy

Tomo-PIV	Tomographic PIV
TR-PIV	Time-Resolved Particle Image Velocimetry
UV	Ultraviolet
VR	Virtualized Reality

Symbols

I	Pixel intensity
x, y, z	Spatial coordinate
θ, φ	Angular coordinate
λ	Wavelength
$w_{i,j}$	Weighting coefficient for MART reconstruction
C_{df}	Center of the plenoptic disc feature (in pixel plane)
D_{df}	Diameter of the plenoptic disc feature (in pixel plane)
p_c	Center of point-like feature
f_m	Focal length of the main lens
$f_{\#}, f/\#$	f-number
p_m	Effective size of the main lens
S_o	Distance from the object focal plane to the main lens
S_i	Distance from the main lens to the image focal plane
M	Magnification of the plenoptic camera, $M = -S_o/S_i$
γ	Tilted-angle, the angle between main lens plane and MLA plane
α	The angle between main lens plane and focal plane, $\tan\alpha = -M\tan\gamma$
f_l	Focal length of the lenslet
p_l	Lenslet pitch
p_p	Pixel pitch
O	Ideal optical center (of the main lens)
O_m	Intersection point between optical axis and MLA plane
O_p	Perpendicular projected point of O_m on sensor plane
A	Aperture center
A_1	Perpendicular projected point of A on MLA plane
A_m	Size of aperture
P	Object point, a point light source
Q	Converging image point
Q_1	Perpendicular projected point of Q on MLA plane

R	Intersection point of the MLA plane and the line that passes through A and Q
p_{ci}	Center of the i-th point-like pattern beneath i-th lenslet
h_{ci}	Center of the i-th lenslet on the senor plane
$C_{l(i)}$	Coordinate of the i-th lenslet center in pixel plane
g_{ci}	Center of the i-th aperture projection point beneath i-th lenslet
l_{ci}	Center of the i-th lenslet on MLA plane
p_c	Center of the point-like pattern beneath a lenslet
h_c	Center of a lenslet on the senor plane
g_c	Center of the aperture projection point beneath a lenslet
l_c	Center of a lenslet on MLA plane
C_{EoC}	Center of the EoC formed on MLA plane
D_{EoC}^{major}	Length of the EoC's major axis
D_{EoC}^{minor}	Length of the EoC's minor axis
D_{CoC}	Dimeter of the CoC
$(x, y, z)_o$	Coordinates of an arbitrary point source in the object space
$(x, y, z)_i$	Coordinates of the image of an arbitrary point source $(x, y, z)_o$ in the image space
D_a	Diameter of the main lens
$z_{o,far}$	DOF limit on the far-side of the focal-plane
$z_{o,near}$	DOF limit on the near-side of the focal-plane
$L(s, t, u, v)$	The 4D light-field, where (s, t) and (u, t) respectively define the intersection location of a ray made with two parallel planes
$L(x, y, \theta, \varphi)$	The 4D light-field, where (x, y) and (θ, φ) respectively define the spatial- and angular-coordinates of a ray
f_l	Focal length of the microlens
$d_{i,l}$	Distance from the MLA plane to the sensor plane
s'	A virtual image point
$d_{o,l}$	Distance from the image-plane to the MLA plane
$E(X_j, Y_j, Z_j)$	Intensity of a voxel at (X_j, Y_j, Z_j)
$I(x_i, y_i)$	Pixel intensity of a light-field image
(X_j, Y_j, Z_j)	Coordinates of the j-th voxel
$(x_i, y_i)(i \in N_j)$	Coordinates of the i-th pixel, N_j is the number of object point influential on the pixel
$w_{i,j}$	Weighting coefficient for MART reconstruction
C_{df}	Center of the plenoptic disc feature (in pixeI plane)
D_{df}	Diameter of the plenoptic disc feature (in pixel plane)
y_c	Intersection of a light ray with the main lens, light ray passes lenslet center
p_{l_sensor}	Lenslet pitch (in sensor plane), $p_{l\_\mathrm{sensor}} = p_l \frac{S_i + f_l}{S_i} = \left\lvert C_{l(i)} - C_{l(i-1)} \right\rvert$

$\mathcal{M}_{4\times 3}$	Mapping matrix that relates a spatial point $P(X, Y, Z)$ With its corresponding C_{df}
w_1	Lenslet weighting coefficient
w_2	Pixel weighting coefficient
y_l	Offset of a light ray from the lenslet center
y_m	Intersection of a light ray made with the main lens
y_p	Intersection of a light ray made with the pixel plane
$O - XYZ$	Lens coordinate system (LCS)
$O_p - UV$	2D pixel coordinate system (PCS)
$S_i^{'}$	The Z-direction distance from Q to main lens
Q_d	Distance between Q and Q_1
$\begin{pmatrix} \mu_{up} \\ \sigma_{up} \end{pmatrix}$	The angle that the light ray $\underset{Ql^{up}}{\rightarrow}$ made with MLA plane
γ_x	Sensor rotation angle along X-axis
γ_y	Sensor rotation angle along Y-axis
A^d	Distance AA_1
k	Correction factor of the major/minor axis length of EoC
a_{0-19}	Polynomial coefficients
L_s	Scheimpflug line
Q_{Recon}	Reconstruction quality
$E_0(x, y, z)$	Voxel intensity approximated by a Gaussian distribution with three voxel diameters
$E_1(x, y, z)$	Voxel intensity of the reconstructed particle image
μ	Relaxation factor
ω	Angular frequency
Re	Reynolds number
I_{ref}	Pixel intensity of images captured at wind-off condition
I	The pixel intensity of images captured at wind-on condition
$A(t), B(t)$	Paint calibration coefficients which are normally dependent on local temperature
P_{windon}	Model surface pressure distribution at wind-on condition
P_{ref}	The model surface pressure distribution at wind-off condition
dP/dx	The streamwise pressure gradient
δ_1	The boundary layer displacement thickness
τ_w	The local mean wall shear stress
$(u, v, w), (U, V, W)$	Velocity in three directions
f	Frequency
$u_{\max}$	Maximum velocity encountered during the blowing cycle
τ	Period of the cycle
$\bar{u}$	Mean blowing velocity
U_0	Momentum flow velocity
d	Orifice geometry width
l	Orifice geometry length
Sr	Strouhal number

L_0	Stroke length
D_{so}	Saddle location
$\boldsymbol{D}$	Velocity gradient tensor
λ_{ci}	Swirl strength
c	Blade chord length
c_x	Blade axial chord length
p	Blade pitch
h	Blade span
β_1	Inlet design flow angle
β_2	Outlet design flow angle
Ma	Mach number
Re_c	Reynolds number based on the blade chord length, c
κ_1, κ_2	Principal curvatures
κ_m, κ_G	Mean and Gaussian curvatures
I_i	Radiation intensity of i-th channel
λ_i	Effective wavelength of the i-th channel
A_i	Calibration factor which depends on the wavelength λ_i
T	True Temperature in Eq. (7.1)
C_2	The second radiation constant
$\varepsilon(\lambda_i, T)$	The spectral emissivity of the target at true temperature T
(S_x, S_y)	Lenslet center (relative to optical center)
C_{pk}	Process capability index

Chapter 1
Introduction

Shengxian Shi

Abstract This chapter introduces the basic principle of light-field imaging, which includes the definition of light-field, different methodologies to capture light-field. Such information will lay a foundation for the light-field particle image velocimetry (PIV) camera design as well as light-field particle reconstruction algorithm development. This chapter also provides a brief review on the evolution of PIV, and illustrates how light-field imaging could benefit the development of such technique.

Keywords Light-field · Camera array · Light-field camera · Unfocused light-field camera

The Light-field

Our world is filled with light, rays that incident from all directions. Confined by the limited viewing angle of our binocular vision system, we could only perceive a small portion of light rays. An eye-catching scene could be unfolded if light rays from more directions were captured and displayed, an example of which would be the famous 1999 movie "The Matrix", where views from different angles were displayed sequentially to create an effect of orbiting around a scene, the so-called "Bullet Time" visual effect.

From the photometric point of view, light ray is a carrier of radiant energy, being visible as well as invisible. In theoretically deducing the photometric computation for a luminous surface (i.e. luminosity, illumination, intensity and brightness), Arun Gershun firstly introduced the concept of light-field, which was defined as a collection of three-dimensional light vectors (Gershun 1939). However, if only the luminous intensity is focused, which is the case for most of the optical diagnostic problems, it would be a function of more than three variables. By taking the spatial, angular, wavelength and time information into account, a seven-dimensional function was defined for light rays from every direction. The so-called plenoptic function (the

S. Shi (✉)
School of Mechanical Engineering, Shanghai Jiao Tong University, Shanghai 200240, China
e-mail: kirinshi@sjtu.edu.cn

S. Shi and T. H. New (eds.), *Development and Application of Light-Field Cameras in Fluid Measurements*, https://doi.org/10.1007/978-3-031-01779-7_1

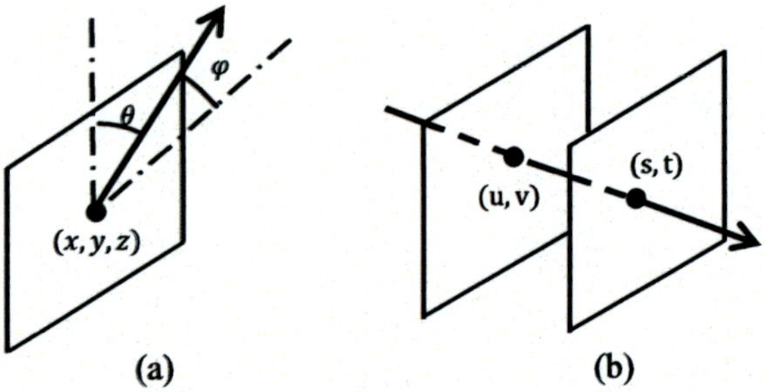

Fig. 1.1 Parametrisation of light-field with **a** five-dimensional variables, and **b** four-dimensional variables

word plenoptic comes from Latin *plenus*, which means complete, full and optic) takes the following form, where (x, y, z) and (θ, φ) denotes the spatial position and angular location of a ray, respectively (Adelson and Bergen 1991).

$$I = I(x, y, z, \theta, \varphi, \lambda, t) \tag{1.1}$$

For PIV-related measurements, normally only single wavelength is involved (e.g. $\lambda = 532$ nm). In addition, the time information is considered by the form of two images (e.g. double-frame exposure PIV) or multiple images (e.g. time-resolved PIV). The above seven-dimensional plenoptic function can therefore be simplified to a five-dimensional one (Eq. 1.2, Fig. 1.1a). Moreover, if only regions outside an object's convex hull are focused (e.g. areas outside PIV tracer particles), and light rays travel in empty space without obstruction (e.g. in transparent medium like air or water), the plenoptic function can be further simplified to a four-dimensional (4D) one, in which a light ray is parameterised by its intersection with two parallel planes (Eq. 1.3, Fig. 1.1b) (Levoy and Hanrahan 1996). Note that light-field is also termed as Lumigraph (Gortler et al. 1996), but the terminology light-field will be applied throughout this book for consistency.

$$I = I(x, y, z, \theta, \varphi) \tag{1.2}$$

$$I = I(u, \mathrm{v}, \mathrm{s}, t,) \tag{1.3}$$

Capturing Light-field

With the light-field concept being defined and parameterised, many devices were developed from the photographic point of view to capture light rays for space outside an object's convex hull, such that effects of flying through scenes or new perspectives could be generated. Following the 4D light-field parameterisation, (Gortler et al. 1996) presented a framework that samples the plenoptic function of static objects with a hand-held camera. By acquiring object images from different viewpoints and estimating camera pose for each frame, light-field of a static object could be constructed and new perspectives can be subsequently rendered from it. To simplify the image acquisition process, a camera can be moved along a track (Katayama et al. 1995), linear translation stages or a spherical gantry (Ashdown 1993) which allows partial or full light-field of a static object to be generated.

Multiple cameras are necessary to capture light-field for dynamic or large scenes, one of such applications is to arrange cameras along a one-dimensional (1D) path to create the effect that a moving object appears frozen in time (Taylor 1996), e.g. Fig. 1.2). Another application is to arrange cameras around a hemisphere or cuboid to capture full light-field for dynamic events, from which full three-dimensional (3D) virtual representations of such events can be reconstructed. The Virtualized Reality (VR) project is a typical example of such application. Pioneered by research group in Carnegie Mellon University, the so-called 3D Dome and 3D Room (Fig. 1.3), which consisted 51 and 49 synchronised cameras respectively, were constructed to simultaneously sample light-field from very dense viewpoints (Rander et al. 1997; Kanade et al. 1998). In addition to the 1D camera array, 3D Dome and 3D Room, two-dimensional (2D) camera arrays were developed to capture full light-field for high performance imaging. One of such examples is the famous Stanford Multiple Camera Array (Fig. 1.4), where 96–100 cameras were arranged in a 2D grid to

Fig. 1.2 The google 16-camera arc

Fig. 1.3 The Carnegie Mellon University 3D room

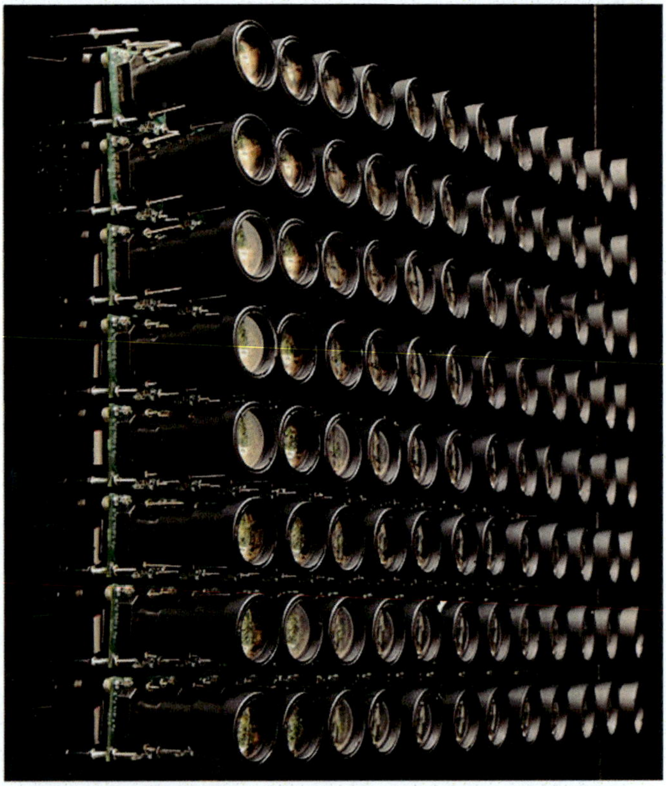

Fig. 1.4 The Stanford multiple camera array

achieve a very high light-field sampling density, thereby eye catching performances such as seeing through occlusion, ultrafast imaging, high dynamic range and high resolution panorama could be realised (Wilburn et al. 2005).

Both the single-moving camera configurations and the multiple camera systems are suitable for wide range (long-baseline) light-field sampling, if smaller range of viewpoints (short-baseline) is focused, the microlens array (MLA)-based camera offers a more compact solution. Such type of camera was termed as plenoptic camera (Adelson and Wang 1992) or light-field camera (Ng 2006), which employs the combination of an MLA and image sensor (e.g. CCD, CMOS) to implement the two-plane parameterisation of 4D plenoptic function (Fig. 1.1b). The terminology light-field camera will be applied throughout this book for consistency. Depending on the separation distance between MLA and image sensor, there are two different types of light-field camera. One is the so-called unfocused light-field camera or light-field camera 1.0, which positions MLA at one focal length of microlens (all microlens have the same focal length) away from the camera sensor plane (Ng et al. 2005; Ng 2006). Put in a simple way, an unfocused light-field camera can be regarded as a miniaturised 2D camera array, where each lenslet and pixels behind it represent a tiny camera that samples light-field of a scene from a particular perspective. Such configuration provides the best angular resolution (determined by the number of pixels beneath each microlens) but moderate spatial resolution (depends on the number of microlens) for light-field sampling. Another configuration, which was termed as focused light-field camera or light-field camera 2.0, positions MLA somewhere between the main lens and image sensor in order to maximise the spatial resolution by sacrificing angular sampling resolution (Lumsdaine and Georgiev 2009). As a new imaging device, commercially available light-field cameras are very few at the moment. Lytro (Fig. 1.5) and VOMMA (Fig. 1.6) provides unfocused light-field camera products, where Lytro aimed for consumer market (Lytro ceased operations in late March 2018) and VOMMA focus on industrial light-field cameras (https://en.wikipedia.org/wiki/Lytro, http://www.vommatec.com/). For focused light-field camera products, Raytrix (Fig. 1.7) is currently the only supplier and its light-field camera consists of microlens with three different focal lengths (https://raytrix.de/).

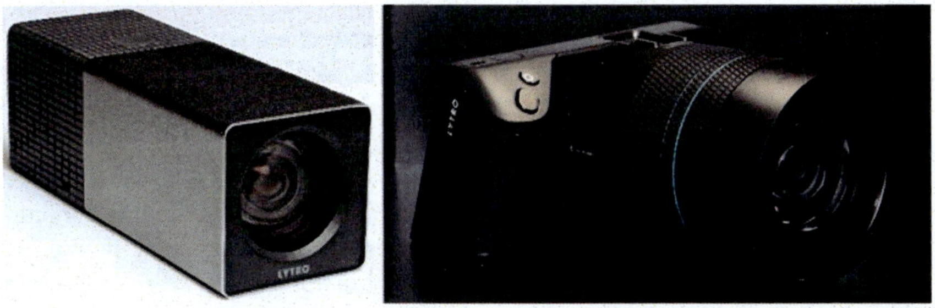

Fig. 1.5 Lytro light-field cameras

Fig. 1.6 VOMMA light-field camera

Fig. 1.7 Raytrix light-field camera

Would Light-field Imaging be Beneficial for PIV?

As a non-intrusive planar velocity measurement technique, PIV has progressed rapidly over the past forty years, and is maturing into a standard fluid diagnostic method which is widely used in many areas such as fundamental fluid mechanics, micro-fluids, bio-fluids, aerodynamics, combustion and turbomachinery (Raffel et al. 2007; Schröder and Willert 2008; Adrian et al. 2011). PIV has its roots in flow visualisation as they both replying on imaging tracer particles to reveal the interior structures of flow fields. One of the very famous examples of the photographic flow visualisation was the one presented by Ludwig Prandtl in 1904 in his work on boundary layer theory, in which the author utilised a film camera to study unsteady separated flows. For quantitatively examining flow structures, speckle photography technique (Archbold et al. 1970; Leendertz 1970) which was originally developed for measuring surface displacement in solid mechanics was extended to fluid mechanics to measure the fluid velocity fields (Barker and Fourney 1977; Dudderar and Simpkins 1977; Grousson and Mallick 1977). This technique is termed as Laser Speckle Velocimetry (LSV), which derives 2D instantaneous velocity field from speckle pattern of small particle photographs that are taken by a film camera. It was later argued that moderate particle density would be desirable for better illumination and easier seeding in large scale or high speed flows, whereby individual particle images instead of speckle patterns were recorded on photographs (Adrian and Yao 1984; Pickering and Halliwell 1984). Before digital imaging technique was available, PIV seeding particles were firstly recorded by negatives, and the photographic recordings were normally digitalised by a slide scanner for subsequent processes.

PIV leap forward in early 1990s when digital cameras and powerful computers were made available for researchers. In contrast to the complex optical interrogation system and cumbersome film developing process, digital PIV is inherently superior than its photographic counterpart as it allows tracer images being captured, viewed at real time and processed without delay (Willert and Gharib 1991; Westerweel 1993). Aiming for better unfolding the complicated fluid phenomena, researchers continued to extend the PIV capability from 2D measurement to two-dimensional three-component (2D-3C) and full volumetric (3D-3C) diagnostics. One of the first efforts was to introduce one additional camera to the traditional two-dimensional PIV (2D-PIV) system, and measure the third velocity component according to stereoscopic imaging principles (Stereo-PIV) (Arroyo and Greated 1991; Prasad and Adrian 1993). A natural extension to Stereo-PIV was to simultaneously measure 2D-3C velocity slices for multiple planes by using a series of scanning laser sheets and a pair of high speed cameras, where the measurable velocity is limited by camera frame rate, laser repetition rate or scanning mirror speed (Brücker 1996; Hori and Sakakibara 2004). Instead of measuring 3D velocity through multiple view geometry, Defocusing Digital PIV (DDPIV) recovers depth information from defocused images which are normally produced by a three-aperture mask. As DDPIV estimates particle 3D coordinate from its triple defocused images, a single camera DDPIV system is limited to flows with very low particle density, and normally a triple-camera arrangement is needed to resolve flow field with satisfactory accuracy (Willert and Gharib 1992; Pereira et al. 2000). One of the truly volumetric velocity measurement techniques is Holographic PIV (HPIV), which records three-dimensional particle displacement by in-line or off-axis holography and subsequently calculates velocity distribution by particle tracking or cross-correlation from reconstructed digital holograms (Hinsch 2002; Katz and Sheng 2010). A significant step forward in the development of 3D velocity measurement techniques is Tomographic PIV (Tomo-PIV), which typically uses four cameras to capture particle images from different viewing angles and reconstructs 3D particle image via multiplicative reconstruction technique (MART) (Elsinga et al. 2006; Scarano 2013). Tomo-PIV has advantages in high spatial resolution as well as relative large measurable volume (measurable range along optical axis is smaller than lateral directions though), and is being widely used in experimental fluid mechanics studies. Another multi-camera 3D velocity measurement technique is synthetic aperture PIV (SAPIV), which is the first research that fuses light-field imaging with experimental fluid mechanics (Belden et al. 2010). In a similar fashion as the Stanford Multiple Camera Array, SAPIV uses a large camera array (normally 8–15 cameras) to capture the light-field image for seeding particles and reconstructs 3D particle image through synthetic aperture refocusing method.

SAPIV can tolerate much higher particle density than Tomo-PIV and its measurable range along optical axis can be on the same order as lateral directions.

Why limiting ourselves in conventional imaging method when seeking for volumetric flow diagnostic solutions? The SAPIV has demonstrated that significant improvement could be achieved via introducing new imaging technique to PIV community. Would it be possible to accomplish 3D flow measurements with lesser cameras or even a single camera without sacrificing performance? Given the ever increasing pixel resolution of the current CMOS image sensor technology (the CCD image sensors are discontinuing when this book compiles), would it be the time to start exploring the possibility of recording more information within one image sensor instead of just light intensities? In addition to large camera arrays, would other light-field recording method like light-field camera be helpful in boosting the PIV technology? Researches have shown positive answers to such questions, and this book will detail recent progress made in light-field camera-based PIV, particularly design and assembly of PIV light-field cameras, light-field calibration and 3D light-field particle reconstruction algorithms as well as applications in experimental fluid mechanics.

After this introduction, the following chapters of this book will introduce: foundational principles and light-field camera designs (Chap. 2), volumetric calibration for light-field cameras (Chap. 3), volumetric flow measurement with single and dual light-field cameras (Chap. 4), 3D surface pressure measurement with single light-field camera (Chap. 5), various applications in volumetric flow measurements for low speed and supersonic flow fields (Chap. 6), and possible development directions for light-field camera-based diagnostics (Chap. 7).

References

Adelson EH, Bergen JR (1991) The plenoptic function and the elements of early vision. Vis Model Group Media Lab Massachusetts Inst Technol

Adelson EH, Wang JYA (1992) Single lens stereo with a plenoptic camera. IEEE Trans Pattern Anal Mach Intell 14:99–106

Adrian L, Adrian RJ, Westerweel J (2011) Particle image velocimetry. Cambridge University Press

Adrian RJ, Yao CS (1984) Development of pulsed laser velocimetry (PLV) for measurement of turbulent flow

Archbold E, Burch JM, Ennos AE (1970) Recording of in-plane surface displacement by double-exposure speckle photography. Opt Acta Int J Opt 17:883–898

Arroyo MP, Greated CA (1991) Stereoscopic particle image velocimetry. Meas Sci Technol. https://doi.org/10.1088/0957-0233/2/12/012

Ashdown I (1993) Near-field photometry: a new approach. J Illum Eng Soc 22:163–180

Barker DB, Fourney ME (1977) Measuring fluid velocities with speckle patterns. Opt Lett 1:135–137

Belden J, Truscott TT, Axiak MC, Techet AH (2010) Three-dimensional synthetic aperture particle image velocimetry. Meas Sci Technol. https://doi.org/10.1088/0957-0233/21/12/125403

Brücker C (1996) 3-D scanning-particle-image-velocimetry: technique and application to a spherical cap wake flow. Appl Sci Res (The Hague) 56:157–179. https://doi.org/10.1007/BF02249379

Dudderar TD, Simpkins PG (1977) Laser speckle photography in a fluid medium. Nature 270:45–47

Elsinga GE, Scarano F, Wieneke B, Van Oudheusden BW (2006) Tomographic particle image velocimetry. Exp Fluids 41:933–947. https://doi.org/10.1007/s00348-006-0212-z

Gershun A (1939) The light field. J Math Phys 18:51–151

Gortler SJ, Grzeszczuk R, Szeliski R, Cohen MF (1996) The lumigraph. In: Proceedings of the 23rd annual conference on computer graphics and interactive techniques, pp 43–54

Grousson R, Mallick S (1977) Study of flow pattern in a fluid by scattered laser light. Appl Opt 16:2334–2336

Hinsch KD (2002) Holographic particle image velocimetry. Meas Sci Technol. https://doi.org/10.1088/0957-0233/13/7/201

Hori T, Sakakibara J (2004) High-speed scanning stereoscopic PIV for 3D vorticity measurement in liquids. Meas Sci Technol. https://doi.org/10.1088/0957-0233/15/6/005

Kanade T, Saito H, Vedula S (1998) The 3D room: Digitizing time-varying 3D events by synchronized multiple video streams. Carnegie Mellon University, The Robotics Institute

Katayama A, Tanaka K, Oshino T, Tamura H (1995) dependent stereoscopic display using interpolation of multiviewpoint images. In: Stereoscopic displays and virtual reality systems II. International society for optics and photonics, pp 11–20

Katz J, Sheng J (2010) Applications of holography in fluid mechanics and particle dynamics. Annu Rev Fluid Mech

Leendertz JA (1970) Interferometric displacement measurement on scattering surfaces utilizing speckle effect. J Phys E 3:214

Levoy M, Hanrahan P (1996) Light field rendering. In: Proceedings of the 23rd annual conference on computer graphics and interactive techniques, pp 31–42

Lumsdaine A, Georgiev T (2009) The focused plenoptic camera. In: 2009 IEEE international conference on computational photography (ICCP). IEEE, pp 1–8

Ng R (2006) Digital light field photography. Stanford Univ

Ng R, Levoy M, Bredif M, et al (2005) Light field photography with a hand-held plenoptic camera

Pereira F, Gharib M, Dabiri D, Modarress D (2000) Defocusing digital particle image velocimetry: a 3-component 3-dimensional DPIV measurement technique. Application to bubbly flows. In: Experiments in fluids

Pickering CJD, Halliwell NA (1984) Laser speckle photography and particle image velocimetry: photographic film noise. Appl Opt. https://doi.org/10.1364/ao.23.002961

Prasad AK, Adrian RJ (1993) Stereoscopic particle image velocimetry applied to liquid flows. Exp Fluids. https://doi.org/10.1007/BF00195595

Raffel M, Willert CE, Wereley ST et al (2007) Particle image velocimetry: a practical guide

Rander P, Narayanan PJ, Kanade T (1997) Virtualized reality: constructing time-varying virtual worlds from real world events. In: Proceedings. visualization'97 (Cat. No. 97CB36155). IEEE, pp 277–283

Scarano F (2013) Tomographic PIV: principles and practice. Meas Sci Technol 24. https://doi.org/10.1088/0957-0233/24/1/012001

Schröder A, Willert CE (2008) Particle image velocimetry: new developments and recent applications. Springer Science & Business Media

Taylor D (1996) Virtual camera movement: The way of the future. Am Cinematogr 77:93–100

Westerweel J (1993) Digital particle image velocimetry—theory and applications PhD Thesis. Delft Univ Technol Netherlands

Wilburn B, Joshi N, Vaish V, et al (2005) High performance imaging using large camera arrays. In: ACM transactions on graphics

Willert CE, Gharib M (1992) Three-dimensional particle imaging with a single camera. Exp Fluids. https://doi.org/10.1007/BF00193880

Willert CE, Gharib M (1991) Digital particle image velocimetry. Exp Fluids 10:181–193

Chapter 2
Light-field Camera Working Principles

Zu Puayen Tan, Brian Thurow, and Shengxian Shi

Abstract This chapter details the working principle of unfocused light-field camera as well as focused light-field camera. It establishes ray propagation model and illustrates synthetic particle image generation for unfocused light-field camera, based on which the design and assembly of embedded-MLA and relayed unfocused light-field camera are introduced. This chapter also provides a procedure for microlens calibration, which is essential for volumetric light-field calibration.

Keywords Light-field imaging · Synthetic light-field image · Light-field camera design · MLA · Microlens calibration

Introduction

Chapter 1 provides an overview of light-field cameras' imaging principles, the different architectures of light-field cameras as well as their hardware designs. Although variations of light-field cameras are numerous, this book is motivated by flow diagnostic applications, where exploiting the compactness, low cost, and simplicity advantages of light-field systems is paramount. Hence, we mainly focus on the MLA-based light-field cameras where these advantages are most pronounced. This is in lieu of other devices that could be considered light-field imagers, such as camera-arrays, view-splitter cameras ("quad-scopes"), and fibre-optically coupled cameras, which are functionally more similar to traditional multi-camera approaches. The basics of conventional imaging will be reviewed to establish a standard set

Z. Puayen Tan
Department of Mechanical Engineering, National Yang Ming Chiao Tung University, Hsinchu 30010, Taiwan
e-mail: tanzu@nctu.edu.tw

B. Thurow
Department of Aerospace Engineering, Auburn University, Auburn, AL 36849, USA
e-mail: thurow@auburn.edu

S. Shi (✉)
School of Mechanical Engineering, Shanghai Jiao Tong University, Shanghai 200240, China
e-mail: kirinshi@sjtu.edu.cn

S. Shi and T. H. New (eds.), *Development and Application of Light-Field Cameras in Fluid Measurements*, https://doi.org/10.1007/978-3-031-01779-7_2

of nomenclatures, and to highlight principles relevant to light-field imaging. This basis will then serve as a foundation for the chapter's core discussions on unfocused light-field camera and, to a lesser extent, focused light-field imaging.

Principles of Conventional Imaging

Figure 2.1 illustrates a conventional camera comprised of an image sensor and a main imaging lens separated by an image distance S_i. The function of the main lens is to collect light rays emitted from points in object space (x_o, y_o, z_o) and to focus them onto equivalent points in the image space (x_i, y_i, z_i). The mapping between any arbitrary point in object space and its equivalent point in image space can ideally be described using the thin lens equation and magnification equation, respectively:

$$\frac{1}{z_o} + \frac{1}{z_i} = \frac{1}{f_m} \tag{2.1}$$

$$M = -\frac{z_i}{z_o} = -\frac{x_i}{x_o} = -\frac{y_i}{y_o} \tag{2.2}$$

where f_m is the focal length of the main lens and M is the magnification. The negative sign reflects the fact that any object is upside down in image space.

In a conventional camera (bottom of Fig. 2.1), an image sensor is positioned at a fixed location in image space and records the intensity of light striking that plane. The location of the image sensor is termed the (s, t) plane with (s, t) representing the 2D image coordinate system and typically expressed in units of pixels. For a fixed sensor location, $z_i = S_i$, there is a unique corresponding plane in object space, termed the nominal focal plane that is located at $z_o = S_o$. Light from all points located along the nominal focal plane (e.g. $(x_o, y_o, z_o)_1$) will be collected and focused onto sharp points on the image space's sensor plane. These object points are thus considered to be "in-focus". The relationship between S_o and S_i is also expressed via the thin lens equation as:

$$\frac{1}{S_o} + \frac{1}{S_i} = \frac{1}{f_m} \tag{2.3}$$

By definition, a system is focused at infinity if the sensor and lens are spaced apart by $S_i = f_m$. Thus, an object such as (x_0, y_0, z_0) at finite distance will necessarily need to be brought into focus by placing the sensor at $S_i > f_m$. The resulting system magnification can then be expressed as:

$$M = -\frac{S_i}{S_o} \tag{2.4}$$

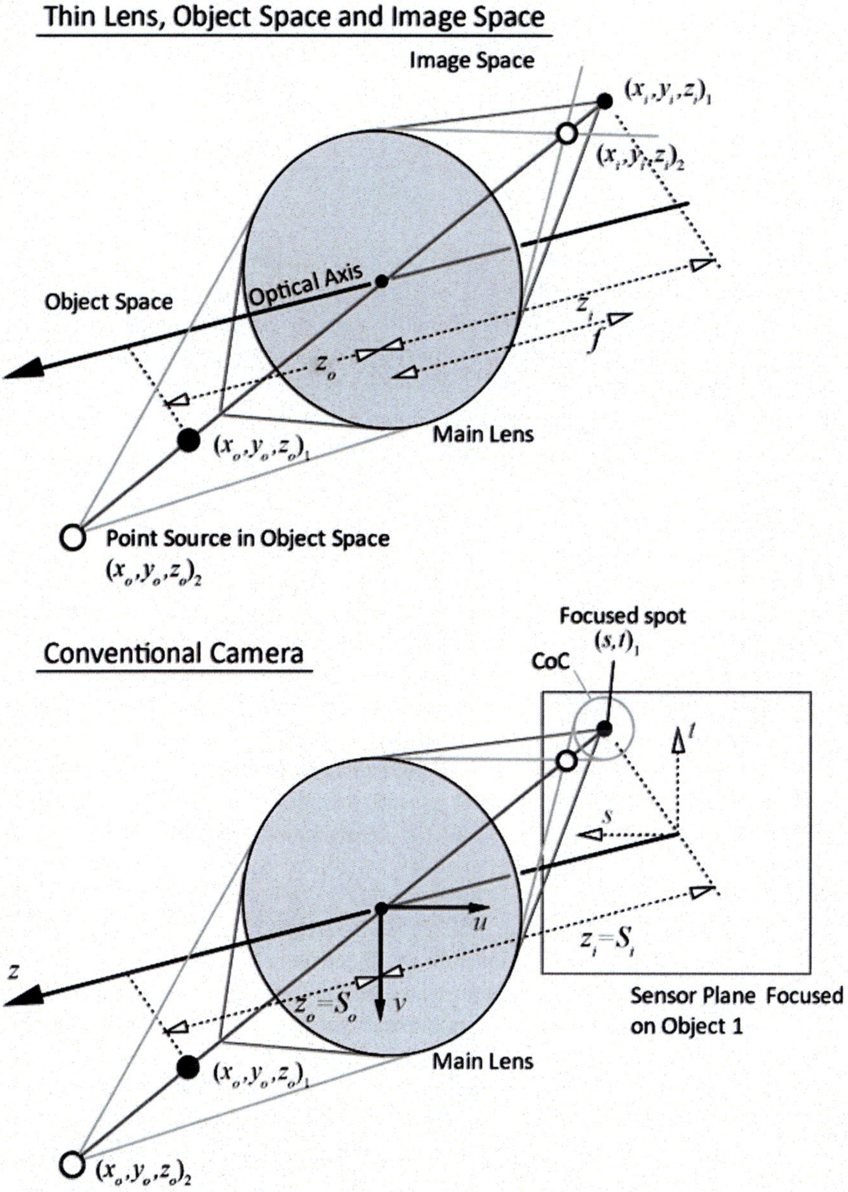

Fig. 2.1 Schematic of rays propagation across object and image spaces (*top*), and the setup of a conventional camera (*bottom*)

On the other hand, the image coordinates (s, t) and world coordinates (x, y, S_o) for objects along the nominal focal plane are described by magnification:

$$s = -Mx_o \tag{2.5}$$

$$t = -My_o \tag{2.6}$$

With a conventional camera, it is not possible to sample the image space arbitrarily along the optical axis (i.e. different z_i) without physically moving the lens or image sensor, commonly referred to as focusing the camera. However, if we are to project the chief ray of arbitrary points in object space (i.e. rays that originate from the point and pass through the main lens' centre) until they intercept the sensor, the corresponding image coordinates of the interception location can still be described by Eqs. (2.5–2.6), which the magnification appropriately replaced by:

$$M_{\text{chief\_ray}} = -\frac{S_i}{z_o} \tag{2.7}$$

$$\frac{1}{z_o} + \frac{1}{S_i} = \frac{1}{f_m} \tag{2.8}$$

where it is noted that both f_m and S_i are fixed, whereas M_{chief_ray} is a variable that depends on object depth, z_o.

A common constraint in the application of conventional imaging for physical measurements is the requirement, or assumption, that the object being imaged is located at the nominal focal plane of the camera system. When an object is not located on that plane ($z_o \neq S_o$), the equations above still hold; however, the depth of the object, z_o, must be known (or determined via image processing algorithms for example) in order to translate dimensions measured in image space, typically in units of pixels, into physical dimensions in object space.

In addition, when an object is not located on the focal plane, the image of a point object will appear to be out of focus, or blurred/smeared into a circular spot known as the circle of confusion (CoC). The degree to which an object is blurred is determined by the diameter of the main imaging lens, p_m, and the object's distance in z away from the focal plane. The relationship between the CoC, z and p_m can be described through simple geometric considerations. For example, consider a point source not located on the nominal focal plane as in Fig. 2.1. Light rays emitted from this point will converge to a point either in front or behind the image sensor. The sensor thus intercepts the ray cone when the cross-section is still a finite circle (the CoC). As the half-angle of the ray cone is dependent upon distance from the focal-plane and p_m, we see that CoC diameter increases in size with distance from $z_o = S_i$ and p_m. This gives rise to the concept of depth-of-field (DoF), which is defined as a range along the optical axis over which the CoC remains within an acceptable limit. This limit is usually taken as the smallest unit of imaging, which for a conventional camera is the size of a single pixel as anything smaller will still be perceived as a point source. The DoF can be expressed as:

$$\text{DOF} = z_{o,\text{far}} - z_{o,\text{near}} \tag{2.9}$$

$$z_{o,\text{far}} = \frac{z_o p_m}{p_m - D_{\text{CoC}}} \tag{2.10}$$

$$z_{o,\text{near}} = \frac{Z_o p_m}{p_m + D_{\text{CoC}}} \tag{2.11}$$

where $z_{o,\text{far}}$ is the limit of the DoF on the far-side of the focal-plane, $z_{o,\text{near}}$ the near-side, and D_{CoC} is the size of the CoC in object-space (typically taken as the pixel-size divided by the system's magnification). As we will see in the ensuing discussions, light-field imaging allows for manipulation of the DoF after an image has been required, often resulting in an extended DoF, which is beneficial for imaging deep 3D volumes whilst maintaining a larger aperture opening for increased light collection.

In anticipation of the discussion of light-field cameras, it is also useful to consider conventional imaging in the context of light-field, which is a conceptual description of light rays as vectors occupying a 3D field. Using Chap. 1's concept, the light-field is described via two-plane parameterisation, where individual light rays propagating from a point source to the sensor can be parameterised in terms of their intersections with two datum planes: (u, v) and (s, t). These planes are most commonly taken as the aperture/main lens plane and sensor plane, respectively, as illustrated in Fig. 2.1. (s, t) is referred to as the spatial-coordinates because objects with different lateral positions along the focal plane will occupy different (s, t), whilst (u, v) is termed the angular-coordinates since the same object's rays will intercept different (u, v) depending on their incident angles on the main lens. Two-plane parameterisation has the advantage that only rays along the same line-of-sight (i.e. photons that will intersect the same point on the sensor) will have the same 4D (u, v, s, t) coordinates. This contrasts with a 3D vector-field description where two or more ray vectors can pass through the same 3D position (x, y, z) at different angles.

Taken together, the full set of light rays collected by a camera system can be referred to as the light-field $L(u, v, s, t)$. In a conventional camera, light is collected at the (s, t) plane, and the process of recording at a point on the sensor can be conceived as an integration of all light rays striking each (s, t) or pixel location:

$$I(s, t) = \iint_{p_m} L(u, v, s, t)\text{d}u\text{d}v \tag{2.12}$$

Due to integration across du and dv, explicit information about the angular content of the light-field is lost in conventional imaging. Visually in Fig. 2.1, the integration

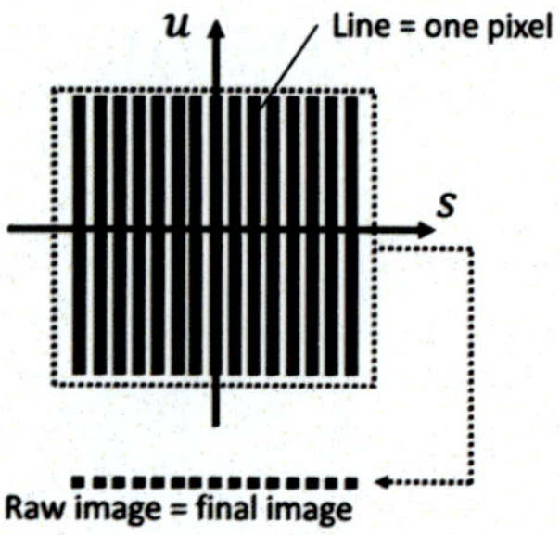

Fig. 2.2 Overview of ray sampling by a conventional camera, plotted on the s-u plane

is equivalent to the sensor's interception of an object's rays where they are focused into a zero-sized point.

Another way to visualise both the sampling and integration process is through spatio-angular sample distribution plots such as Fig. 2.2. The spatial coordinates (s, t) and angular coordinates (u, v) in Fig. 2.2 are remapped as perpendicular axes here, and only the dimensions of s, u are plotted for simplicity. Each thick line segment represents the information collected by a single sensor pixel. A conventional camera's pixels are finely distributed along (s, t), whilst each pixel integrates across the entire range of available (u, v). For instance, the sensor's information is fully devoted to spatial resolution. Hence, in a conventional camera, the final output image (i.e. a 2D array of pixels in (s, t)) is formed by integrating over all (u, v) collected over the main lens aperture. This concept is often used in discussion of light-field imaging (e.g. (Georgiev and Lumsdaine 2010a)) where (u, v) coordinates are sampled in addition to (s, t).

Imaging Principles of Unfocused Light-field Camera

To motivate the reason behind light-field imaging: if the complete light-field of a 3D image space is known, we could technically form images at any arbitrary position and orientation, with arbitrary depth-of-field. And consequently, since the image space is a 3D equivalent of the real object space, this will also allow us to retrieve any three-dimensional measurements on imaged objects. However, as discussed in the previous section, our sampling device (the sensor) is confined to a 2D plane, and thus our recorded data is an incomplete slice of the 3D image space and its 4D light-field. Multi-camera, view-splitter, and fibre-optic approaches attempt to capture more knowledge about the light-field by employing several discrete imaging perspectives, either by using several cameras or separating the views of one camera. Light-field imaging, on the other hand, is a class of imaging principles and device aimed at multiplexing rich 4D light-field information onto an otherwise 2D sensor, which effectively redistributes our sample points in the light-field from a 2D (s, t) plane into the fuller 4D volume. Emphasis was placed on "redistribute" because the total amount of information remains a conserved quantity, and we will discuss that a

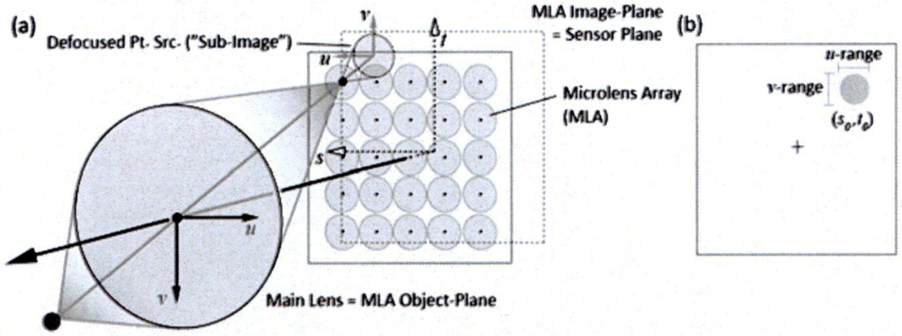

Fig. 2.3 **a** Schematic of ray-capturing in a unfocused light-field camera for an object at the main lens' nominal focal plane (distance S_o from mains lens), **b** example of how an in-focus point source would appear

pixel of information devoted to capturing (u, v) information is a pixel fewer in (s, t) resolution. Nonetheless, the knowledge of the 4D light-field, even if only sparsely samples, allows creative image rendering techniques such as perspective-shifting, refocusing, and even 3D metrology to be performed on a single 4D light-field image.

Our discussion begins with the original type of light-field camera architecture, known as plenoptic 1.0 or unfocused light-field camera. The basic imaging principles behind unfocused light-field camera are introduced via Fig. 2.3. The unfocused light-field camera is nearly identical to convention camera except for the addition of an MLA between the sensor and the main lens. This MLA, consisting of sub-millimetre microlenses packed into a rectangular or hexagonal pattern, is placed precisely at S_i behind the main lens; i.e. at the image plane (s, t) corresponding to the system's nominal focal plane. The function of each microlens is to collect the light arriving at that particular (s, t) location and to distribute the light onto the pixels behind the microlens according to the incident angle at which the light arrived. To accomplish this, the image sensor is precisely positioned one microlens focal length (f_l) behind the MLA, a gap usually on the order of $o(0.05 - 5\text{ mm})$. In this configuration, the image sensor records an array of sub-images, whereby each microlens forms an image that is nominally focused at "infinity" relative to the microlens. At the relative scale of the components, "infinity" is treated as being coincident with the main lens aperture-plane (i.e. the (u, v) plane). Thus, each pixel in a sub-image collects light passing through a small portion of the main lens aperture, where the pixel location in the sub-image corresponds to a discrete (u, v) coordinate on the aperture plane. Collectively, the array of sub-images recorded by a unfocused light-field camera represents a 4D sampling of the light field, with pieces of the (u, v) dimensions multiplexed within each (s, t) point. In practice, the raw 2D image data recorded by the image sensor is typically restructured into a 4D array with dimensions corresponding to the (u, v, s, t) coordinate system, which helps to facilitate the application of light-field image processing algorithms based on the two-plane parameterisation of the light field.

The fundamental design of an unfocused light-field camera can be further understood by considering Fig. 2.4, which schematically illustrates the image formation process for a point source at different locations in object space. In Fig. 2.4a, the point

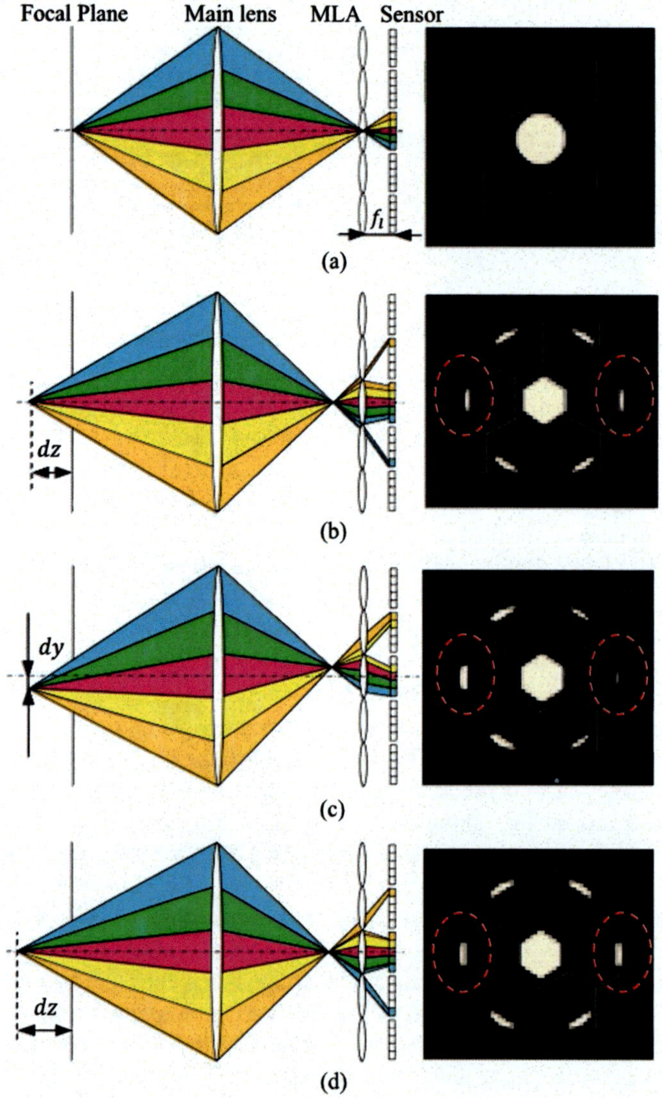

Fig. 2.4 Illustration of light-ray sampling for an object on the near-side and far-side of the main lens focal-plane. The green dash blocks denote the projection of the borders of the microlens on the sensor plane. Light-field images generated by ray-tracing simulation: **a** light source on the focal plane centre, **b** light source offset from the focal plane by $dz = 0.385$ mm, **c** light source offset by $dz = 0.385$ mm and $dy = 0.0055$ mm, and finally **d** light source offset by $dz = 0.405$ mm

source is located at the focal plane of the main lens. The main lens collects the light emitted by the point source and focuses it onto a single microlens at the centre of the MLA. The microlens directs the incident light onto the image sensor, which is shown here with a resolution of 5 pixels per microlens (pixel-to-microlens ratio, PMR = 5). A sample image is also included where it can be observed that the microlens image is a filled-in circle corresponding to the shape of the main lens aperture. Each pixel on the image sensor records a portion of the angular content collected by the main lens and passing through that microlens with colour used to indicate the subset of light rays associated with each pixel. Figure 2.4b, c illustrates a point source that has been shifted forward along and laterally from the optical axis, respectively. As the point source is moved off the main lens nominal focal plane, it forms an increasingly large CoC that eventually spans across multiple microlenses. As before, the pixels illuminated underneath each microlens record the angular information associated with the subset of light rays striking that particular microlens. When the point source is on the far-side of the focal plane, the focused image is in front of the MLA, and the rays are diverging as they approach the MLA. Thus, the rays land on the outer edges of the outer circle of microlenses. Conversely, when the point source is on the near-side of the focal plane, the rays are converging and will hit pixels on the inner edge of the outer circle of microlenses. When the point source is moved laterally, the proportion of rays occupying the outer circle of microlenses also shift accordingly, similar to that expected in a conventional imaging system. Lastly, a comparison of Fig. 2.4b, d emphasises the fact that unfocused light-field cameras only sample the 4D light-field up to some limited resolutions. Point sources that are displaced by a small amount in depth may appear identical in the raw image.

Figure 2.5 presents a raw image of a conventional scene recorded with a unfocused light-field camera and includes a zoomed in view showing the individual microlens images. When viewed in its entirety, the image appears similar to what might be expected with a conventional camera. This is due to the fact that the microlens images are vanishingly small compared to the overall scale of the image such that the angular content captured in each microlens image is too small to discern and is naturally integrated out by the human eye. Thus, the image appears equivalent to an image acquired with a conventional camera equipped with an image sensor with

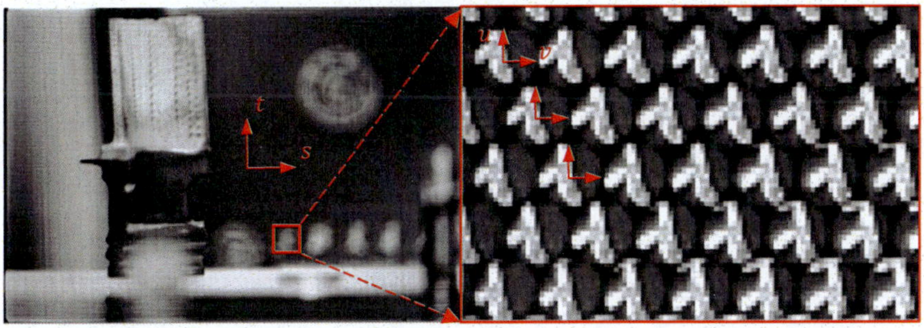

Fig. 2.5 Example of a raw unfocused light-field image

pixels the size of a microlens. The location of focal plane is determined by the focus setting of the main lens with a DoF that is controlled by the main lens aperture p_m. In practice, unlike conventional camera with arbitrary p_m, the unfocused light-field camera's p_m is a dependent variable. The raw output from a unfocused light-field camera is often used to adjust the focus and p_m of the main lens. Similar to Fig. 2.3, Fig. 2.5 shows the (s, t) coordinate system of the raw image with an origin that is fixed at the centre of the image sensor, which is assumed to coincide with the optical axis. The (u, v) coordinate system, on the other hand, does not have a fixed location on the image sensor, but rather is defined separately for each microlens with its origin set at the centre point of each microlens sub-image. Consequently, since the sub-image is mapped to the (u, v) aperture plane, the radius of each sub-image is proportional to p_m. One method for determining the origin of (u, v) in each sub-image is to reduce the aperture of the main lens, which reduces the extent of each microlens image to nearly a point that corresponds to the centre of the aperture. A centroid-finding or similar algorithm can then identify the location of these points to sub-pixel accuracy.

The critical role that the main lens aperture plays in the set-up of an unfocused light-field camera is visualised in Fig. 2.6. Microlens images for three different settings of the main lens aperture are shown. When the aperture is set too small, there is a large gap between neighbouring sub-images resulting in a large underutilisation of the image sensor's resolution. On the other hand, when the aperture is set too large, the sub-images overlap causing ambiguity in the measurements. The ideal condition is to have microlens images that nearly touch, but do not overlap. This condition, recognised by (Ng 2006), is known as f-number matching (or "F-match"). By considering similar triangle relationship for rays on the edge of the main lens/sub-image:

$$\frac{p_m}{S_i} = \frac{p_l}{d_{i,l}} = \frac{p_l}{f_l} \tag{2.13}$$

where the additional subscript l refers to parameters connected to the lenslet whilst subscript m refers to parameters connected to the main lens. As a reminder, the microlens image distance $(d_{i,l})$ is set to f_l by definition of an unfocused light-field camera's architecture. If we define the image-side f-number on the main lens as $((f/\#)_m = f_m/p_m)$ and that of the microlens as $(f/\#)_l = d_{i,l}/p_l = f_l/p_l$, and taking into account the relationship between f_m and the system magnification (i.e. Eqs. 2.5 and 2.6), the ideal f-number setting of the main lens can be related to the fixed f-number of the microlenses using Eq. 2.14

$$(f/\#)_m = (f/\#)_l/(1 - M) \tag{2.14}$$

As an example, in many applications, 1:1 imaging magnification on the main lens such that $x_i = x_o$ and $M = -1$ is quite common. Under this condition, an MLA with $f/4$ microlenses will require a main imaging lens with an $f/2$ aperture

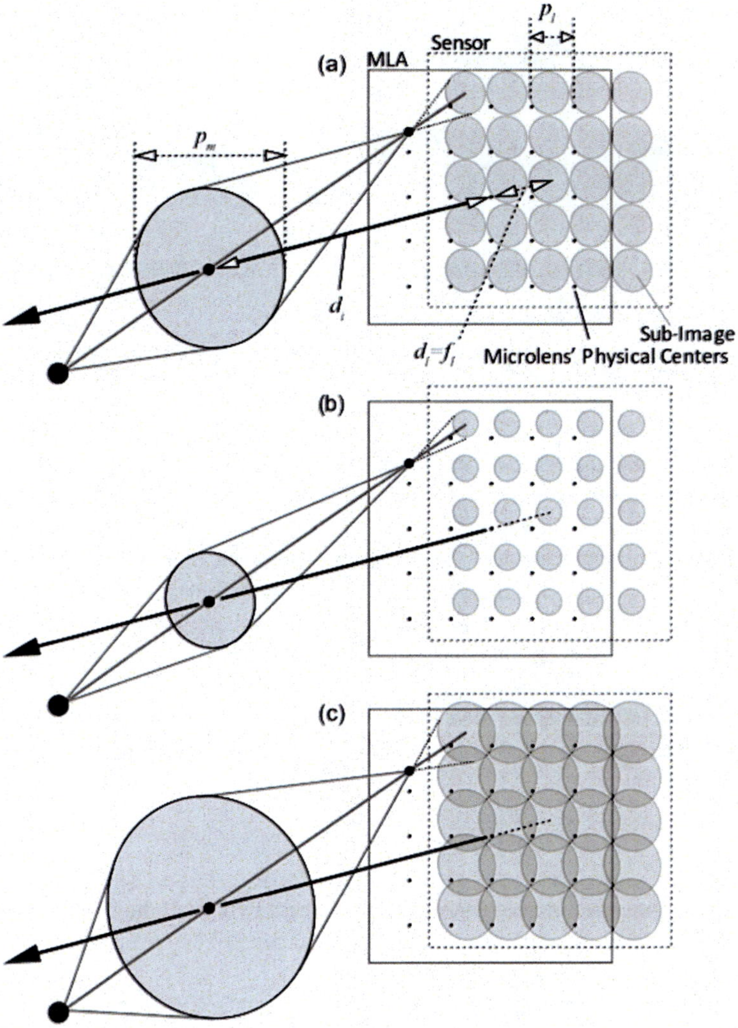

Fig. 2.6 Illustration of the effects of aperture diameter (p_m) on the sub-image pattern. **a** Correctly adjusted p_m, creating a condition known as "F-match". **b** Undersized p_m. **c** Oversized p_m

in order to satisfy this condition. In addition, the working distance for a lens set at a 1:1 magnification will be twice the focal length of the lens (i.e. 1:1 magnification with a 100 mm focal length lens will be achieved when both the object and image distance are set to 200 mm). The availability of large aperture (low $f/\#$) lenses with the desired focal length or working distance often places a practical constraint on the design of unfocused light-field systems. Notably, the angular resolution in degrees-per-pixel remains identical across all three cases in Fig. 2.6; however, the ranges of (u, v) defined by the sub-image diameter differ. And since (u, v) range is equivalent

(a) (b)

Fig. 2.7 Example of new perspective images generated from the raw light-field image shown in Fig. 2.5

to parallax baseline size in stereo-camera 3D imaging, obtaining large but non-overlapping sub-images is crucial for obtaining high depth precision in unfocused light-field-based 3D metrology.

Figure 2.7 illustrates one of the primary functions enabled by the capture of light-field image data with an unfocused light-field camera: the ability to generate unique perspective views from a single snapshot. As described previously, each pixel in a microlens image collects the light that passes through a small portion of the main lens aperture. A perspective view image can be synthetically generated by constructing an image consisting of only the subset of light collected at the same relative pixel location in each microlens image. Put in equation form, this is equivalent to interrogating $L(u, v, s, t)$ over a range (s, t) values for a fixed value of $(u, v) = (u_0, v_0)$:

$$I_{u_0,v_0}(s,t) = L(s,t,u = u_0, v = v_0) \tag{2.15}$$

By changing the value of (u_0,v_0), one can create views of the world as seen from different positions on the main lens aperture. It can also be observed in Fig. 2.7 that the DoF of an individual perspective view is relatively large. This is due to the fact that the effective aperture size for this image is relatively small and equivalent to p_m/PMR, or more appropriately, the number of pixels spread across the diameter of a sub-image. The spatial resolution of Fig. 2.7 is fundamentally determined by the number of microlenses in the MLA and is a direct consequence of the fact that the number and spacing of each microlens determine the discretisation of (s, t). This latter point represents the most significant trade off associated with unfocused light-field cameras. Namely, for an image sensor of a given size, the spatial resolution of images generated from the data will experience a reduction in spatial resolution by a factor of PMR.

The perspective view image generation process can alternatively be understood by considering Fig. 2.8. This figure presents the data captured by an unfocused light-field camera in the form of a spatio-angular sample distribution plot. Here, the large rectangular boxes represent the information captured by each microlens. The horizontal width of the box along the (s, t) axis corresponds to the size of

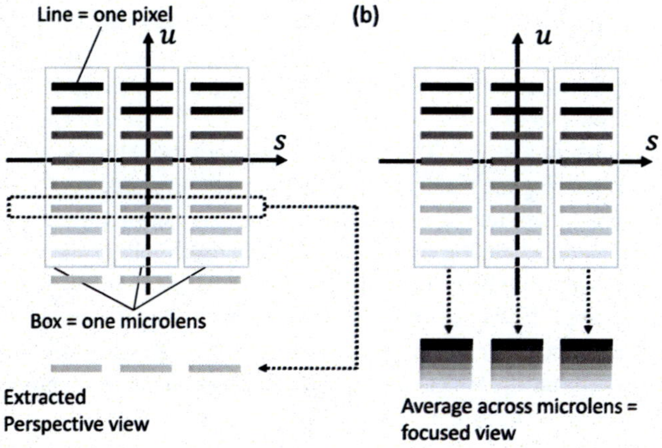

Fig. 2.8 **a** Generation of perspective-views from rays encoded in the *s*-*u* plane. **b** Generation of focused-view for an unfocused light-field camera

the microlens and the vertical height along the (u, v) axis corresponds to the size of the main lens aperture. Pixels on the image sensor are represented by thick, line segments, which are contained within each microlens and uniformly distributed along the (u, v) axis. The generation of a perspective view image can be visualised as extraction of only those pixels located at a fixed (u, v) location, shown in Fig. 2.8a using a dotted line. Figure 2.8b also illustrates how one would generate a "focused" image by integrating the pixels values across a range of (u, v) locations for a fixed (s, t) position. This is equivalent to reconstructing light intensity across the entire lens aperture by summing up the discretised components. An example of a focused image is shown in Fig. 2.9. Due to the integration of all (u, v) content across the full aperture, the focused image has a shallow DoF characteristic of conventional imaging with a large aperture lens.

If the focused image is generated by summing across the (u, v) dimension in the raw image, the resulting focal plane (i.e. where the object appears sharp) will

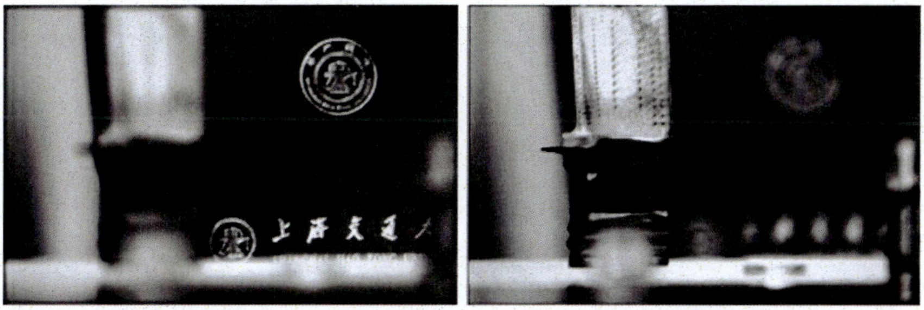

(a) (b)

Fig. 2.9 Example of refocused images generated from the raw light-field image shown in Fig. 2.5

be coincident with the main lens' nominal focal plane as set during image capture. However, perhaps the most well-known function associated with light-field cameras is the ability to computationally refocus an image to a different focal plane after the image is taken. An example of computationally refocused images at two different planes is shown in Fig. 2.9. The process of computational refocusing can be understood by considering the fact that the image space is a volume containing various rays captured by the main lens from the object space. A nominal focal plane only exists in so far as that is where the sensor sampled the light-field of the image space. For a conventional camera where (u, v) data is absent, images can only be rendered where rays are captured. However, with the full set of (u, v, s, t) data as captured by the microlens array, angular information (u, v) can be used to project rays within each pixel to any arbitrary planes along the optical axis. Image formation (i.e. integrating across rays across (u, v) for each (s, t)) is then performed at these virtual image planes, whose rendered image will contain a new focal plane that is associated with the virtual image plane's position.

The processes of image rendering on the nominal focal plane and refocusing are contrasted in Fig. 2.10, which is shown in s and u dimensions only for clarity. The new focal plane is located a distance $S_i' = \alpha S_i$ from the aperture plane. For $\alpha < 1$, this is equivalent to moving the image sensor closer to the main lens whereas for $\alpha > 1$, this is equivalent to moving the image sensor away from the main lens. Figure 2.10b illustrates the fundamentals of this process by considering a virtual image point (s') located on the new focal plane. By definition, rendering of a focused image about this virtual point s' is accomplished by integrating all rays which originate from the aperture u-plane (i.e. rays of all angles captured by the main lens) and which pass through s'. The remaining procedure then concerns where on the raw image to retrieve the samples needed to perform this integration. Through geometric optics, Fig. 2.10b shows that a particular ray sample (u_0, s_0') can in fact be retrieved by propagating this ray all the way to the MLA plane where the raw data is captured. Where $(s_0' - u_0)$ is the lateral distance between the ray's intercepts on the u- and

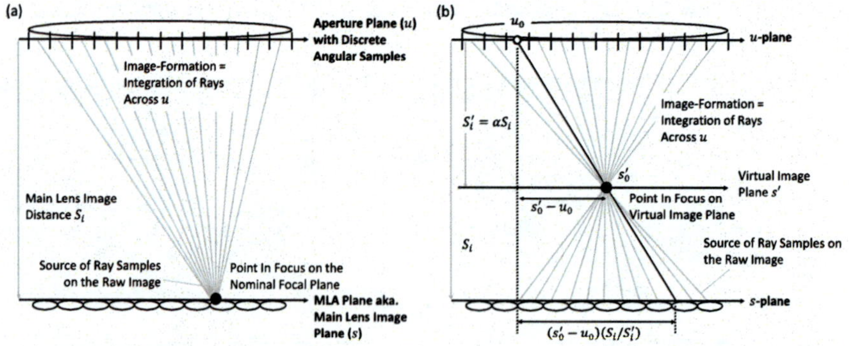

Fig. 2.10 Schematic of **a** generating an image focused at the nominal focal plane and **b** refocusing an image

s-planes, $(s'_0 - u_0)(S_i/S'_i) = (s'_0 - u_0)(1/\alpha)$ denotes its intercept with the MLA plane. Consequently, in terms of the full 4D light-field $L(u, v, s, t)$ on the MLA plane, a refocused image can be generated by integrating together all of the light rays that pass through the main lens aperture plane and crossing through the point (s', t'):

$$I(s', t') = \iint_{P_m} L\left(u\left(1 - \frac{1}{\alpha}\right) + \frac{s'}{\alpha}, v\left(1 - \frac{1}{\alpha}\right) + \frac{t'}{\alpha}, u, v\right) dudv \quad (2.16)$$

In comparing the refocused images in Fig. 2.9 to the perspective views in Fig. 2.7, the DoF of the refocused images is markedly shallower than the perspective view images. This is a direct consequence of the fact that the integration is being carried out over the full aperture of the main lens. Similar to the perspective view image, the effective spatial resolution of a refocused image is fundamentally constrained by the size and number of microlenses, thus representing a significant trade-off with unfocused light-field cameras. This trade-off is not as problematic, however, when one considers the ability to arbitrarily set the focus of the image after it's been taken.

The precise details of the image processing algorithms and numerical methods utilised to transform raw unfocused 2D image data into a 4D light-field matrix, and the subsequent generation of perspective view and computationally refocused images from this data are beyond the scope of this chapter. It is noted, however, that these details can have a significant impact on the overall image quality and the potential appearance of several different types of image artifacts. Readers interested in learning more about such implementations are directed to the open source software package, Light-field Imaging Toolkit (Bolan et al. 2016), which contains a graphical user interface, a suite of Matlab functions, and sample data that can serve as a convenient starting point for exploring some of these issues.

Imaging Principles of Focused Light-field Camera

Since the application of focused light-field camera in flow diagnostics is limited and nascent, our treatment of this architecture in this chapter will be brief and only touches upon several key concepts.

Following the original work on unfocused light-field camera, a later architecture pioneered by (Lumsdaine and Georgiev 2009) is the so-called "plenoptic 2.0" or "focused light-field camera". The microlenses are no longer coincident with the main lens' image-plane, but are rather positioned with respect to the sensor such that they are themselves focused on the image-plane. In this manner, the MLA acts as an array of microscopic relay lenses replicating small cropped copies of a nominal image-plane onto the sensor.

Using the same example of a point source at the main-lens' nominal focal-plane, the resulting ray propagations in a focused light-field system are shown in Fig. 2.11a. An array of barely contacting sub-images is also produced (Fig. 2.11b) in

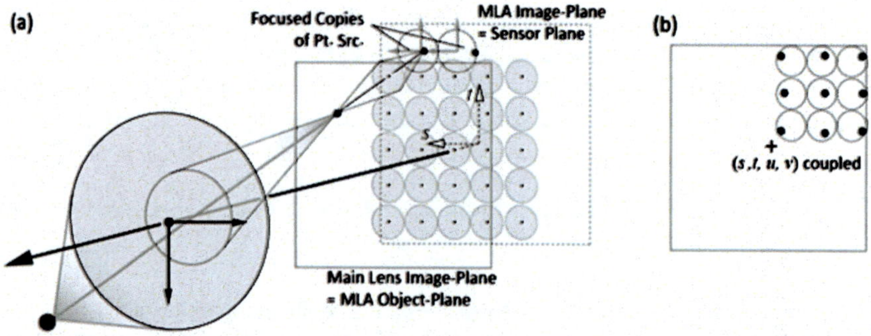

Fig. 2.11 **a** Schematic of ray-capturing in a focused light-field camera, **b** example of how an in-focus point source would appear

the same manner as unfocused light-field camera. However, unlike unfocused light-field systems, since the MLA's image is now focused on the main-lens' nominal image-plane, the point source appears in focus. Dependent upon the point source's distance with respect to the nominal focal-plane, copies of the point image may occur across multiple sub-images with different degrees of position shifts (i.e. "disparity" in the sense of multi-camera stereo-vision) with respect to the microlens' centre. This depth-dependent disparity allows for 3D information to be gleaned in focused light-field systems. The illustration shown in Fig. 2.11a is that of a Keplerian configuration where the MLA is behind the main lens image-plane, whereas a Galilean configuration also exists where the MLA intercepts the rays before the image-plane (Lumsdaine and Georgiev 2009). Since the ratio of object distance from the image-plane to the MLA ($d_{o,l}$) and the image distance from MLA to sensor ($d_{i,l}$) defines the degree of magnification with which the MLA relays the image-plane onto the sensor, different trade-off of angular to spatial resolutions can be achieved by changing the $d_{o,l}/d_{i,l}$ ratio.

Whilst unfocused light-field camera contains a dense distribution of pixels in (u, v) at fixed (s, t), the imaging principles in focused light-field camera mean that (u, v) and (s, t) are coupled within a microlens. On the s-u plot, as illustrated in Fig. 2.12, this means that the samples of each microlens are sheared in the s-u domain. Herein is the primary distinction of unfocused and focused light-field cameras. The former contains dense samples in the (u, v) dimension but is relatively sparse in (s, t), whilst the latter may contain different distributions of samples across (s, t) and (u, v) depending on the degree of "shear" of each light gray boxes. In both conventional and unfocused light-field cameras the distributions of samples are locked to hardware, whereas in focused light-field camera the "shearing" of samples is directly controlled by tuning the $d_{o,l}/d_{i,l}$ ratio—a key engineering advantage of focused light-field camera.

Since (s, t) and (u, v) are coupled in focused light-field camera, a "pure" perspective-view is very challenging to render. Figure 2.12a shows a naïve way of rendering perspective-view from focused light-field camera data: an arbitrary band

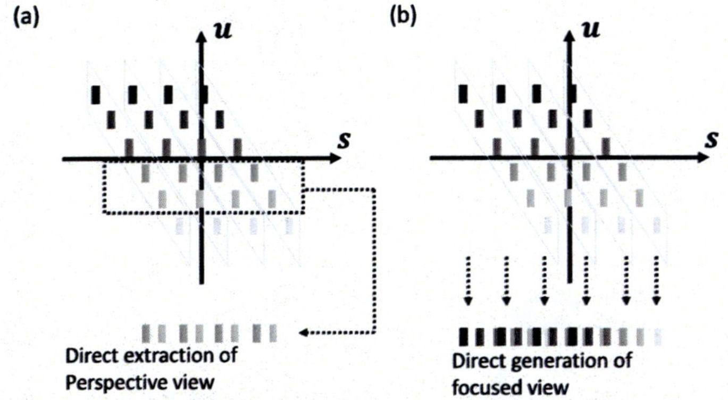

Fig. 2.12 **a** Generation of perspective-views from rays encoded in the *s*-*u* plane. **b** Generation of focused-view for a focused light-field camera

of (u, v) is extracted and stitched together. This represents extraction of multi-pixel patches from each sub-image for tiling. Though naïvely feasible, this rendering has higher (u, v) uncertainty and typically also contains tiling artifacts as (u, v) values zipper across the image. Details of artifact minimisation are beyond the current scope, but can be found written by (Georgiev and Lumsdaine 2010a, b). Refocusing can also be performed in focused light-field camera (Fig. 2.12b), typically to much higher spatial resolution than unfocused light-field camera. Though depending on the sophistication of the rendering algorithm, tiling artefacts can again emerge. Typically, disparities between sub-images are computed using cross-correlation as part of the decoding step in focused light-field camera, which indirectly produces a depth map of the scene, such that smooth tiling using the correct patch-sizes can be accomplished (Georgiev and Lumsdaine 2010a, b).

Synthetic Image Generation and Point Source Response

Though much regarding light-field camera design can be accomplished via first-order approximates and models of optics such as the thin lens equation as discussed in the sections above, there comes a point in verifying the design's performance where synthetic light-field images have to be generated and evaluated. For flow diagnostics, this often involves simulating the intended camera design's response to point sources that resemble flow-tracing particles (e.g. for PIV applications). This commonly also involves generating perspective views from these synthetic images, and processing these perspective views through a separate tomography or triangulation algorithms to reconstruct 3D particle fields (as is the case for 3D PIV applications). In such manners, the impacts of different camera parameters on reconstruction performance or any other intended functions can be rapidly evaluated empirically, instead of relying on physics-driven optical models, which can quickly become intractable.

Synthetic image generation fundamentally relies on ray-tracing through the light-field camera, commonly using the method of Ray Transfer Matrix (RTM) analysis, which will be detailed below. The RTM procedure is then combined with a Monte Carlo procedure that initiates a rich set of rays emanating from some specified objects to the main lens. Once propagated through the entire light-field camera, the collection of rays intersect and are recorded on the simulated sensor to form the synthetic image. For a PIV-related light-field recording, rays will originate from a point-like object resembling a seeding particle.

The two-plane parameterisation is slightly modified from $L(u, v, s, t)$ to $L(x, y, \theta, \varphi)$ for the RTM analysis, where (x, y) and (θ, φ) define the intersection location and angle of a ray made with a plane perpendicular to the camera optical axis, respectively. Light ray propagation inside an unfocused light-field camera is illustrated in Fig. 2.13 (shown in 1D for simplicity). The RTM operators that describe ray propagation from the point object P to the sensor plane are described in Eqs. 2.17–2.21. In these equations, S_o is the object distance, S_i is the image distance, f_m is the focal length of the main lens, f_l is the focal length of the microlens, p_m is the aperture of the main lens, p_l is the microlens pitch, p_p is the pixel pitch, S_x and S_y are the shift of microlens centre from the main optical axis.

Free-space propagation is first used to describe ray propagation from particle P having position $(x, y, z = z_o)$ to the main lens:

$$\begin{pmatrix} x' \\ y' \\ \theta' \\ \varphi' \end{pmatrix} = \begin{pmatrix} 1 & 0 & z_o & 0 \\ 0 & 1 & 0 & z_o \\ 0 & 0 & 1 & 0 \\ 0 & 0 & 0 & 1 \end{pmatrix} \begin{pmatrix} x \\ y \\ \theta \\ \varphi \end{pmatrix} \tag{2.17}$$

In this case (θ, φ) can be randomly seeded by the Monte Carlo algorithm, with enough cases to populate the entire range of anticipated angle. Rays travelling through the main lens experiences no positional shift, but are refracted in angle using the operator:

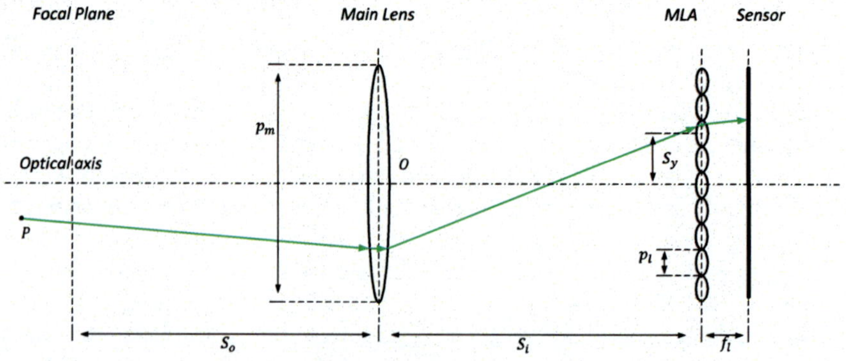

Fig. 2.13 Light ray path inside an unfocused light-field camera

$$\begin{pmatrix} x' \\ y' \\ \theta' \\ \varphi' \end{pmatrix} = \begin{pmatrix} 1 & 0 & 0 & 0 \\ 0 & 1 & 0 & 0 \\ -1/f_m & 0 & 1 & 0 \\ 0 & -1/f_m & 0 & 1 \end{pmatrix} \begin{pmatrix} x \\ y \\ \theta \\ \varphi \end{pmatrix} \tag{2.18}$$

A second free-space operator describes propagation from the main lens to the MLA:

$$\begin{pmatrix} x' \\ y' \\ \theta' \\ \varphi' \end{pmatrix} = \begin{pmatrix} 1 & 0 & S_i & 0 \\ 0 & 1 & 0 & S_i \\ 0 & 0 & 1 & 0 \\ 0 & 0 & 0 & 1 \end{pmatrix} \begin{pmatrix} x \\ y \\ \theta \\ \varphi \end{pmatrix} \tag{2.19}$$

Followed by refraction through the MLA:

$$\begin{pmatrix} x' \\ y' \\ \theta' \\ \varphi' \end{pmatrix} = \begin{pmatrix} 1 & 0 & 0 & 0 \\ 0 & 1 & 0 & 0 \\ -1/f_l & 0 & 1 & 0 \\ 0 & -1/f_l & 0 & 1 \end{pmatrix} \begin{pmatrix} x \\ y \\ \theta \\ \varphi \end{pmatrix} + \begin{pmatrix} 0 \\ 0 \\ S_x/f_l \\ S_y/f_l \end{pmatrix} \tag{2.20}$$

In Eq. (2.20), it is necessary to know which microlens the ray intersects, because refraction is with respect to the local microlens optical axis, rather than the main lens' optical axis. Finally, the ray travels a short distance (f_l) from the MLA to CCD via free space propagation:

$$\begin{pmatrix} x' \\ y' \\ \theta' \\ \varphi' \end{pmatrix} = \begin{pmatrix} 1 & 0 & f_l & 0 \\ 0 & 1 & 0 & f_l \\ 0 & 0 & 1 & 0 \\ 0 & 0 & 0 & 1 \end{pmatrix} \begin{pmatrix} x \\ y \\ \theta \\ \varphi \end{pmatrix} \tag{2.21}$$

With the above equations, light-field image of a point light source or a seeding particle can be simulated by randomly emanating a large number of rays, which brim the main lens aperture. It is found that five million rays are sufficient to generate a statistically meaningful light-field image (Shi et al. 2016). As an example, the previously discussed Fig. 2.4 shows four synthetic light-field images for a point source located at four different positions. Such analysis can be very useful in developing a new light-field system, as metrics such as resolution, depth precision, and others can be estimated directly from the synthesised raw images (Deem et al. 2016; Shi et al. 2016).

Light-field Camera Design

The challenge of designing a light-field camera's hardware is associated with how to precisely align the MLA's image-plane onto the camera sensor—in addition to balancing MLA and sensor performance parameters. This challenge is twofold: (i) physically accommodating the small $o(0.05 - 5\text{ mm})$ clearance between MLA and sensor, and (ii) the extremely short focal length of the MLA making it orders of magnitude more sensitive to misalignment than the main lens. In implementation, dimensional differences between unfocused and focused light-field cameras are minute, since the MLA on focused light-field camera is moved only a fraction of a focal length relative to unfocused light-field camera's. However, focused light-field cameras may also require additional means of adjusting the MLA image-to-object distance ratio by users. Hardware types for both unfocused and focused light-field cameras can be categorised into (i) embedded-MLA cameras and (ii) relayed-MLA cameras. Below, we discuss aspects of the MLA first, followed by each type of camera architectures.

Microlens Array

Aside from lens sizes, the first issue of MLA design involves the packing pattern: whether to lay out the microlenses in a simple rectangular grid, a hexagonal honeycomb grid, or more exotic combinations. In one of the landmark work in light-field camera design, (Ng 2006) reported that hexagonal microlens can greatly increase pixel usage efficiency over a rectangular microlens. The reason is illustrated via Fig. 2.14: a hexagonal packing requires individual microlens to be fabricated in a hexagonal shape, whilst a rectangular packing a square shape. Although individual microlens can be fabricated into either hexagonal or square form with pixels filing the entire area underneath, the circular aperture of the main-lens necessarily means that the actual sub-images can only be inscribed circles. In this sense, the hexagonal microlens has significantly less unused areas between the circular sub-image and the microlens' border. Note, in both cases, all the rays that fall onto the MLA will be captured on a pixel, and therefore light capturing efficiency is the same. However, the light will be distributed across more pixels in the case of the hexagonal MLA. One distinct advantage of the rectangular MLA is that sampled rays can be directly assigned onto the rectangular pixel-grid of a digital image, whereas hexagonal-to-rectangular interpolation is required for the hexagonal MLA. The latter often results in noticeable interpolation artifacts, especially on supposedly straight edges. Thus, the advantages and disadvantages of each MLA packing pattern are worth consideration during camera design.

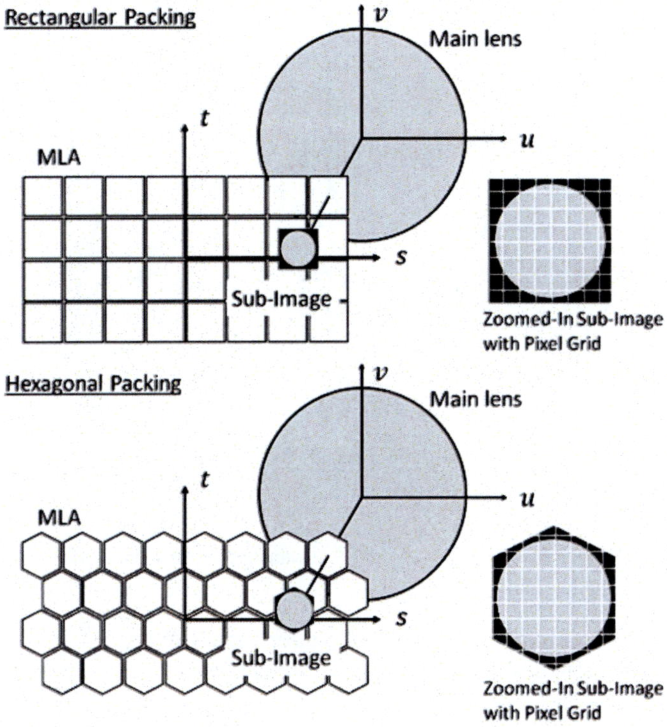

Fig. 2.14 A comparison of rectangular versus hexagonally-packed MLA and their pixel-use efficiency

Embedded-MLA Light-field Camera

A generic schematic of an embedded-MLA light-field camera is shown in Fig. 2.15, next to a conventional camera, which often acts as a basis for light-field modifications. Two notable features in a regular camera pertain to designing embedded-MLA systems: firstly, for Nikon F-mount cameras (a very common lens format in large sensor research cameras) a standard distance of 46.5 mm separates the sensor from the lens' attachment flange, called the flange focal length (FFL). Space within the FFL is highly constrained and it is typically difficult to place retrofit components (e.g. the MLA) here without modifying the camera. Secondly, a protective glass is customarily installed at least several millimetres in front of the sensor to protect the chip from dust and foreign objects. As the name suggest, embedded-MLA light-field camera has its MLA mounted directly to the sensor, where the MLA's image-plane coincides with the sensor. This brings up the immediate design challenge of clearance and positioning accuracy. For unfocused light-field camera the MLA-to-sensor distance is one microlens focal length, and that of focused light-field camera is very similar. Most microlenses' focal lengths are on the order of $o(0.05 - 5\text{ mm})$. Thus,

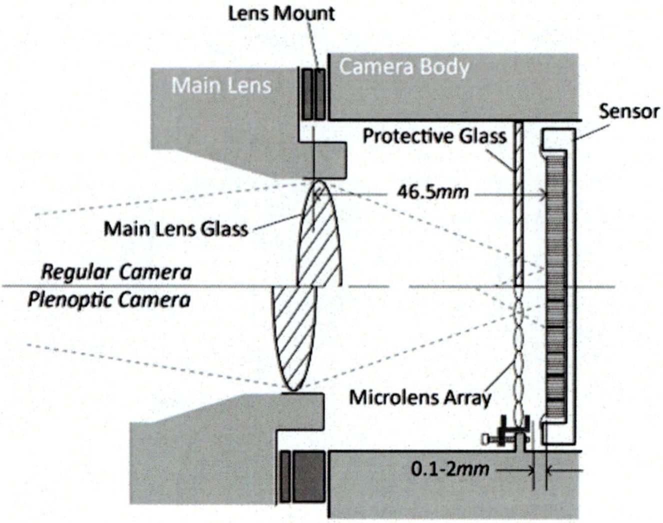

Fig. 2.15 Schematic of regular versus light-field camera with embedded MLA

the protective glass must be removed to accommodate the MLA. Thereafter, the MLA can be mounted via low profile adjustment screws or permanently glued with precise spacers on the sensor.

Following MLA installation, the main lens' flange must also be extended forward to account for the shift in image-plane position from the sensor to the MLA. Additionally, as introduced in Sect. 2.2 Imaging Principles, since F-matching is critical to produce contacting but non-overlapping sub-images, a means to precisely control the main lens F-number (i.e. aperture) is necessary. Many modern commercial main lenses are designed without manual aperture control. Any control is accomplished via a lever-pin that extends from the lens' built-in aperture ring into a matching camera's aperture-control mechanism. Older lenses tend to have manual aperture adjustment rings, but controls are usually only available in discrete steps that are too coarse for F-matching. Thus, a separate aperture control mechanism is usually attached between the main lens and the camera flange to achieve precise F-matching. The impact of this mechanism's thickness on the image distance must be accounted for.

The case-study of a research-grade embedded-MLA light-field camera is presented here to illustrate the engineering decisions and process behind developing such hardware. The hardware, shown in Fig. 2.16, is an in-house unfocused light-field camera that was developed at Shanghai Jiao Tong University (Shi et al. 2016) specifically for PIV measurements. There are two main considerations in building the camera: achieving maximum sampling resolution for the light-field and maintaining short inter-frame time to allow for PIV cross-correlation with small time-increments. The first consideration translates to a high-resolution CCD or CMOS sensor. Although many digital single-lens reflex cameras (DSLR) that satisfy the resolution requirement are commercially available (e.g. Mamiya Phase One iXR

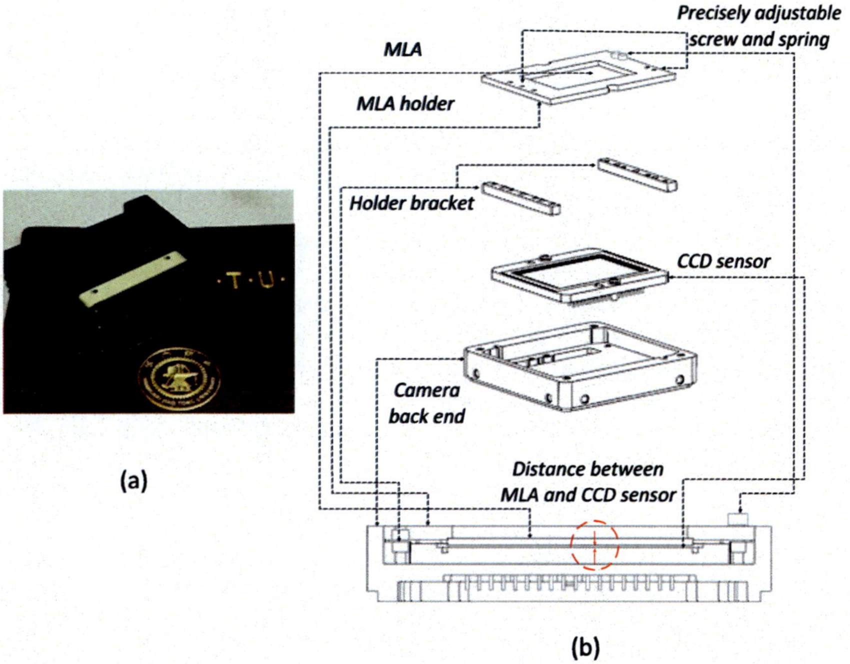

Fig. 2.16 **a** In-house assembled light-field camera; **b** exploded schematic of camera and MLA mounting apparatus

with resolution of 80 Megapixel, and Canon EOS 5D Mark III with resolution of 50.6 Megapixel), they are inherently unsuitable for PIV applications due to their large time interval between consecutive frames. Thus, specialised research-grade cameras are required. When the camera in Fig. 2.16 was built in December 2014, the highest resolution PIV camera available was the Imperx B6640, which uses a KAI-29050 CCD sensor with a resolution of 6600 × 4400 pixels. In addition to resolution and inter-frame time, the sensor's physical dimension is often a secondary factor that must be considered. Larger sensors are generally desirable, because for a given magnification, lens diameter, and working distance (i.e. given parallax baseline, which defines depth uncertainty), the larger sensor can capture a larger field-of-view in object space, thus encompassing a larger volume in the case of PIV application. For a given pixel size and PMR, a larger sensor is also less demanding on the degree of MLA micro-fabrication. Thus, in addition to its high resolution and short inter-frame time, the B6640's full-frame sensor dimension of 36 × 24 mm makes it an ideal foundation for development.

The next step in designing the in-house light-field camera is the selection of a suitable MLA. An unfocused light-field system typically biases the MLA selection towards smaller microlens diameters and a hexagonal packing, allowing more microlenses to be packed within the sensor's surface. This is indispensable as one microlens approximately equates to one rendered image pixel in unfocused light-field camera. However, arbitrarily small microlens cannot be adopted, because smaller diameter equates to fewer pixels per microlens, equivalent to lower angular resolutions and shallower DOFs (as discussed in Sect. 2.2.2). For PIV applications where the intended subjects are deep volumes, high spatial resolution and high angular resolution are both important. The in-house light-field camera was optimised through synthetic images to have 14 pixels per microlens diameter, ideal for PIV volumes on the order of 35–70 mm in lateral dimension (i.e. $M = 0.5$–1.0), and 25–50 mm in depth (Shi et al. 2016). Having selected the microlens diameter, the focal length (or F-number) has to be optimised. The choice of F-number is typically based on consideration of intended working distance and available main-lenses. In this case a Tamron 60 mm macro lens was considered for achieving $M = 1.0$, as it has an exceptionally large aperture of F-number = 2 (i.e. a large parallax baseline for high depth precision). To achieve necessary F-matching whilst fulling utilising the Tamron main-lens' aperture, the MLA needed an F-number of 4. In culmination, the chosen MLA is a customised component with hexagonal packing factor, a focal length of 0.308 mm, F-number of 4, and a resolution of 408×314 microlenses across the full-frame sensor.

Given the core components of sensor and MLA, the next consideration involves mechanical accommodations, i.e. how to precisely position the MLA at one focal length away from the CCD plane, which is 0.308 mm for the current configuration. This minimal separation requires removing the protective cover glass from the CCD sensor, as the thickness of which is on the order of millimetres. As such, the assembling work was carried out in a Class 1000 clean room to avoid the CCD and MLA being contaminated by dusts in the air. The design of the apparatus for suspending the MLA precisely in front of the sensor is shown schematically in Fig. 2.16. It consists of a milled MLA holder, within which seats the fragile MLA glass. Two brackets straddle between the holder and edges of the sensor. Spring-loaded fine-adjustment screws are used to achieve precise micrometer-level MLA positioning in 3D translation and orientation. Raw images achieved on this light-field camera were shown in Fig. 2.5, whilst the perspective views were shown in Fig. 2.7 and refocused images in Fig. 2.9.

Other examples of embedded-MLA light-field cameras intended for PIV applications can be found in (Lynch 2011; Fahringer and Thurow 2012, 2018; Fahringer et al. 2015), which also discuss the data-processing workflow from imaging to PIV volume reconstruction in details. On the other hand, examples of commercially-available light-field cameras with embedded-MLA are shown in Figs. 1.5–1.7. These include the very early Lytro (Butler 2012; MARKUS 2012; Cho et al. 2013; Dansereau et al. 2013), then the Lytro Illum (Pierce 2014), and later the VOMMA light-field cameras (Ma et al. 2019; Zhao et al. 2020) both of which are unfocused light-field architecture. Both models were not optimised for flow diagnostics per se but represent important steps in commercialising the technology. Parts of the Lytro cameras such as MLA were occasionally re-purposed for research prototypes (Dansereau et al. 2017). The

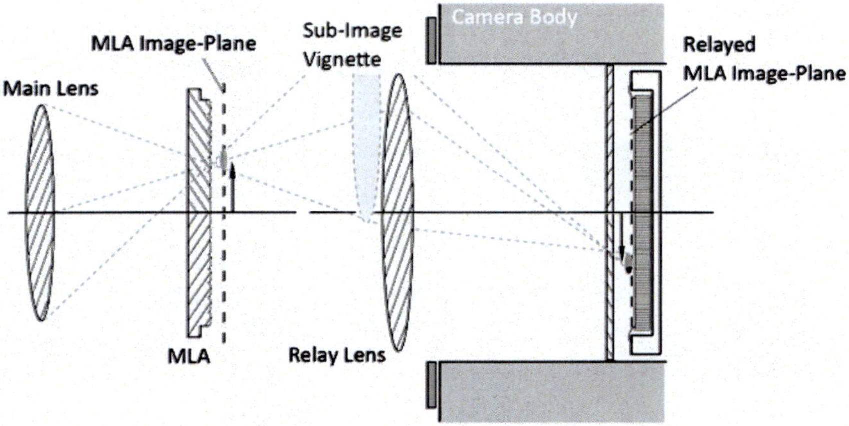

Fig. 2.17 Basic schematic of the relayed unfocused light-field camera

third example is Raytrix's focused light-field cameras, which were used in general 3D metrology as well as flow velocimetry.

Relayed-MLA Light-field Camera

The second type of light-field hardware, the relayed-MLA camera, is shown in Fig. 2.17. As the name suggests, the MLA is now mounted externally to the camera body, and a relay lens is used to re-project the MLA's image-plane onto the sensor. This configuration is motivated by the strong preference to avoid risky modifications to cameras for MLA installation (e.g. the removal of protective glass and risk of damaging image sensor), especially if the application is prototyping instead of production. As a side benefit, relayed-MLA designs are also inherently convertible between regular and light-field imaging without cleanroom operation—an important convenience for high-cost cameras shared between groups with different usages. Finally, whilst the afore-mentioned benefits are merely conveniences, relaying the MLA becomes a necessity when the image-sensing component contains protective glasses or front windows that simply cannot be removed or thinned down. This scenario includes any attempt to couple MLA with an image intensifier whose sensing surface is inside an evacuated tube with a thick front window.

We illustrate the details and performance quirks of relayed-MLA camera via the case study of "DragonEye", an unfocused light-field system designed at Auburn University (Tan et al. 2019, 2020) and shown in Fig. 2.18a, b. DragonEye was conceived from the outset as a flexible light-field system capable of both kHz-rate acquisition for time-resolved 3D PIV, and intensified kHz-rate imaging for flow diagnostics applications that extend into the UV or IR ranges (e.g. OH-PLIF, FLEET and pyrometry). Hence, a relayed MLA is both preferred for 3D PIV and required for

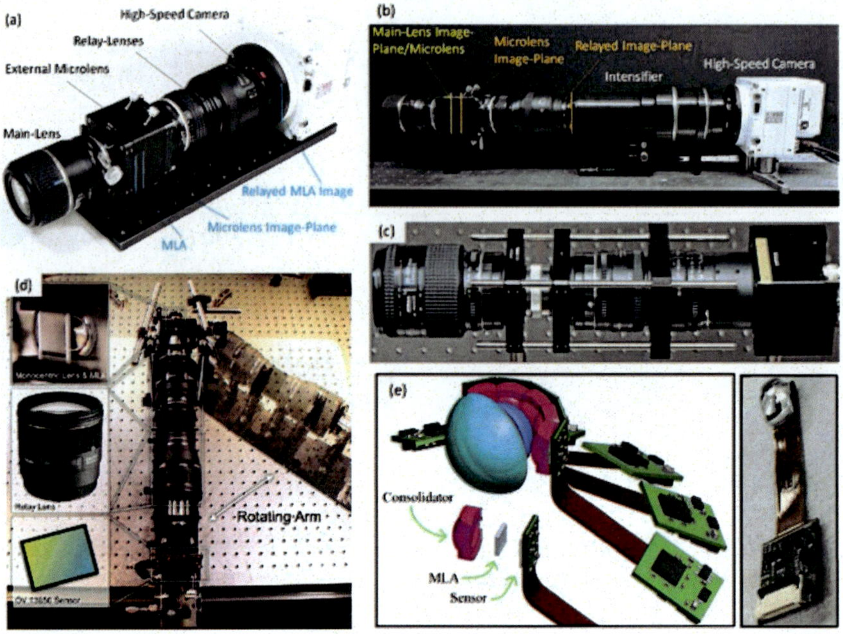

Fig. 2.18 Examples of light-field cameras with relayed MLA: **a** Auburn University's DragonEye, **b** DragonEye installed on intensifier for low-light/UV imaging, **c** (Liu et al. 2017)'s caged light-field system which was used both for macro and microscopic imaging, **d** Dansereau's prototype monocentric-lens light-field camera (Dansereau et al. 2017), **e** a more matured version of the prototype where MLA mounting has been switched to embedded type. Adapted with permission from (Schuster et al. 2019) © The Optical Society

intensified imaging. Like many prior systems for PIV, unfocused light-field camera was adopted for its higher angular resolution and simple (s, t)-(u, v) sampling principles. This necessitated both a high-resolution MLA and a high-resolution sensor, as was the case for embedded-MLA camera. Following from the heritage of (Fahringer and Thurow 2018), DragonEye adopted an existing design of custom MLA with hexagonal microlens packing pattern, 0.077 μm microlens diameter, and an F-number of 4 (i.e. specifications that are also very similar to Fig. 2.16's MLA). Notably, this custom MLA is also of quartz construction as to be UV-transparent.

Although higher sensor resolution is almost always desired, acquisition speed practically limited the maximum resolution which can be obtained. For instance, many camera sensors are limited in data throughput, and higher resolutions come at the sacrifice of frame rates. At the time of design in 2018, resolution-to-speed trade-off limited DragonEye's camera choice to two models: Vision Research's Phantom VEO4k-990 and IDT's OS10-4 k, both featuring approximately $4000 \times 2400 px$ resolution and 1 kHz frame-rate. Ultimately, the VEO4k was chosen for DragonEye due to its larger sensor size of 27.6×15.5 mm compared to the OS10's 17.9×11.2 mm. On the one hand larger sensor provided better light sensitivity, but more critically on the other hand, it provided higher final image resolution when paired with the existing MLA (i.e. a larger sensor spans a greater portion of the 35 mm-sized MLA, making use of a higher total number of microlenses).

With the 0.077 μm MLA and VEO4k, the mechanical layout of DragonEye is as Fig. 2.18a shows, which in turn was inspired by Fig. 2.18c (Liu et al. 2017). An external cage/box-like structure holds the MLA. Internally, this structure contains either a focusing tube for axial MLA adjustment or multiple adjustor screws for both tilt and travel to facilitate micrometer-level focusing. Relay lenses are attached on one end to the MLA structure, and on the other end directly to the camera's lens mount, thereby cutting out any light leakage and constraining all components to a common optical axis. In Fig. 2.18c's design, called the "caged light-field camera", 60 mm optical cages and rails provided system rigidity. On DragonEye, rigidity was achieved via an optical base plate, which allowed for larger MLA and lenses.

The use of a large sensor and MLA became a challenge for DragonEye at this point. As Fig. 2.17 illustrates, rays exiting the microlenses resemble directed ray-cones defined by the main lens aperture (see gray dashed lines). Unlike a uniformly diffused light source that is identical when imaged from any angle, any part of the ray-cone that is not intercepted by the relay lens are lost to the camera. For a light-field system, this shows up as clipped sub-images on the edges of raw image (Fig. 2.19b) and as severe vignette when rendered into perspective-view or refocused images. Thus, the challenge essentially became one of finding a sufficiently large-aperture relay lens to encompass the 27.6 × 15.5 mm MLA/sensor. Implicitly, it was also a challenge of minimising relay lens working distance via shorter focal lengths, in order to minimise any further cone divergence between the MLA the relay's front element. These requirements are difficult to achieve simultaneously, since smaller focal lengths equate to smaller lens diameter for a given F-number, and F-numbers are limited to 1.2–1.8 on practical lenses.

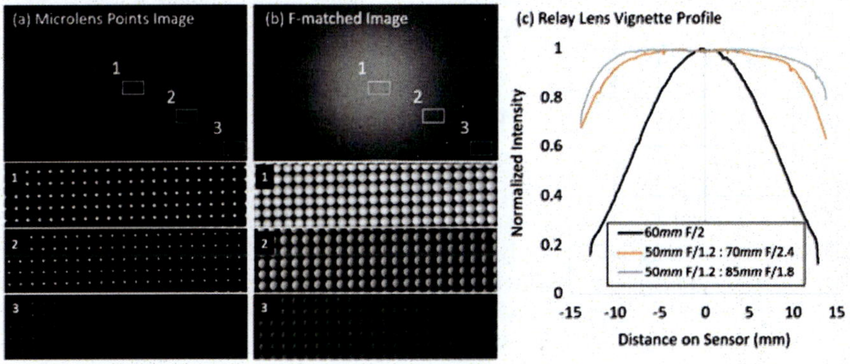

Fig. 2.19 **a** Microlens dots (relay PSF) image, **b** F-matched image of a white background, **c** relay lens vignette profile, altogether illustrating the challenges of relayed-MLA systems: aberration and vignette

The relay lens' working distance and aperture requirement touches upon the second challenge of designing relayed-MLA systems: optical aberration. Most lenses have spherical aberrations and a non-zero field-curvature (i.e. a spherical focal-plane). On a normal lens these aberrations are noticeable but often acceptable, but on the scale of microlens where the relay's focusing plane must match to the microlens' sub-millimetre focal lengths, tolerances to aberrations are exceptionally low. As noted by (Drazic et al. 2012), an insufficiently low curvature may cause parts of the image to lose their light-field properties altogether. To further exacerbate design difficulties, aberrations typically worsen with decreasing F-number, placing the aberration requirement in direct conflict with the vignette requirement.

DragonEye's initial prototype used the Tamron 60 mm F/2 macro lens as relay, which had the advantage of unity magnification at the closest working distance of just 100 mm. F-number is also very low for its class, and the lens is optimised for a low field curvature in macro imaging. Figure 2.19a is a raw image sample captured with the 60 mm F/2 arrangement when the main lens aperture has been closed down to a minimum (i.e. each microlens which is focused on the main lens aperture plane effectively sees a point source, thus characterising the relay system's PSF distribution across the image). Evidently, the 60 mm F/2 lens sharply relayed the microlens points in the centre 2/3 of the image (insets 1–2). Inset 3 contains noticeable tangential astigmatism, but not to a debilitating extent. In spite of its short working distance and low F-number, vignette was an issue using the 60 mm F/2 lens. Figure 2.19b shows that when the main lens is F-matched and imaging a white surface, microlenses in the centre (inset 1) contain complete sub-images that touch without overlapping, whilst insets 2 and 3 contain increasingly clipped off sub-images. The intensity distribution of Fig. 2.19b across a horizontal slice in the centre of image is plotted in Fig. 2.19c. For the 60 mm F/2 case, vignette occurs immediately in off-centre regions.

The vignette with 60 mm F/2 was deemed unacceptable for DragonEye. A final compromised solution was found through a dual-lens relay setup: a Nikon 50 mm F/1.2 positioned flange-forward, coupled to a Pentax 70 mm F/2.4 mounted flange-aft. Mounted flange-forward, the first lens could be positioned within 46.5 mm of the MLA image plane (i.e. FFL distance). At the same time, being a double-Gauss design, the Nikon 50 mm F/1.2 has one of the largest pieces of flange-side glass of any lens, maximising ray capture. The 50 mm Nikon lens positioned at FFL from the MLA image plane projects this plane to infinity, whilst a second lens set to focused on infinity is needed to reform the final image onto the sensor. A symmetric 50 mm:50 mm setup did not mitigate vignette sufficiently. Thus, DragonEye's second lens employed a longer focal length to magnify the MLA image by × 1.4 (i.e. zooming past the vignette). This also reduces the system's effective MLA resolution by × 1.4 (i.e. a smaller portion of the MLA was captured), which was deemed acceptable. The final vignette profile is shown in Fig. 2.19c. A 50 mm:85 mm configuration was also attempted, but did not reduce vignette sufficiently as to justify its resolution loss. Finally, a set of refocused images taken with the DragonEye in its final configuration is shown in Fig. 2.20 to provide an example of its performance. Though lower in

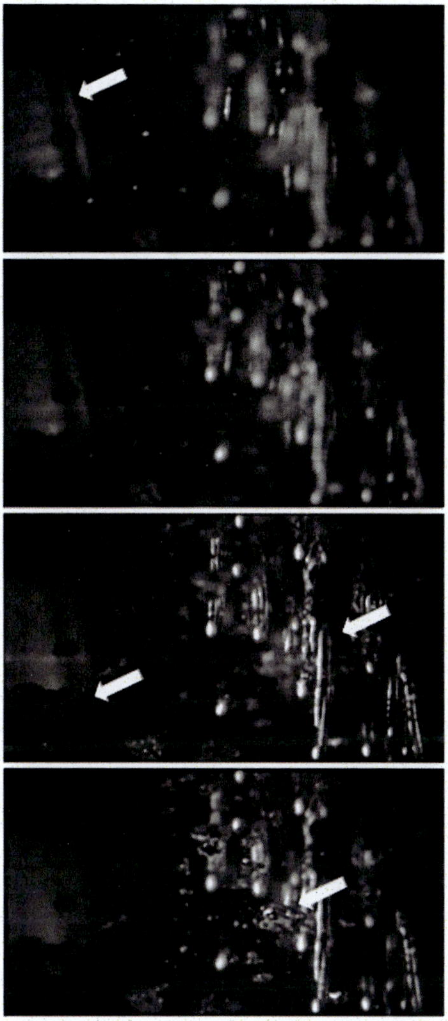

Fig. 2.20 Refocused images of ball-bearing plunging into a transparent water tank, captured with DragonEye 50 mm: 70 mm configuration

resolution compared to the embedded-MLA camera discussed in Fig. 2.16, DragonEye is presently the only kHz-rate light-field camera beside Raytrix's new R21 model (the latter based on embedded-MLA and focused light-field camera).

Figure 2.18 contains examples of several other relayed-MLA cameras. This includes in Fig. 2.18c of (Liu et al. 2017)'s "caged light-field" design that matches an MLA to an Imperx type low-speed camera. Interestingly, this device has been applied to macro-scale flow imaging (Qi et al. 2019) as well as microscopic imaging (Song et al. 2019). Figure 2.18e is an example of relayed-MLA being used in prototyping, prior to transition into a more compact and higher optical quality embedded-MLA design on the matured prototype. Here, the motivation was the use of a simple monocentric lens in lieu of complex fish-eye design. However, the monocentric lens has a spherical image-plane which typically required curved sensors. In this case, light-field refocusing is used to map a curved image-plane to a flat sensor. The matured prototype eschews the relay and directly glues the MLA onto the sensor.

Microlens Calibration

Camera calibration for multi-camera 3D measurements and conventional imaging involves two parts: establishing the camera's extrinsic parameters (i.e. the camera's 3D position and orientation in a defined real world) and intrinsic parameters (i.e. the light projection relationship from a camera-centric 3D world, such as the object space, onto the camera's 2D sensor space). Similar principles apply for light-field cameras, since each rendered perspective view can be conceived as an independent camera, and a collection of perspective views is identical to images from a set of closely-bundled conventional cameras. Since these calibrations relate to image processing and is not strictly relevant to light-field camera design, they will not be discussed here, except in passing reference to well-established techniques in light-field flow diagnostics, such as (Hall et al. 2018)'s polynomial and volumetric dewarping technique, (Hall et al. 2018)'s direct light-field calibration, and the plenoptic disk feature and point-like feature-based methods (O'brien et al. 2018; Shi et al. 2019; Zhao et al. 2020; Zhao and Shi 2021), which will be detailed in Chap. 3. Unique solutions presented in these techniques can be largely traced to the challenge of establishing calibration for $o(100)$ possible perspective views, compared to a conventional system's 2–4 views per measurement setup.

Instead of calibrating for extrinsic and intrinsic parameters, this section deals with how to adjust the MLA for imaging, and subsequently establishing the necessary MLA-related parameters to operate the light-field camera once it is assembled. These include the steps (shown in Fig. 2.21) of (1) focusing the MLA, (2) microlens centre-finding, and (3) the afore-mentioned F-matching.

Step 1 is used to adjust the MLA until it is located at f_l from the sensor to some tolerance (for an unfocused light-field camera system). Since adjustments may be performed via small screws on the MLA holder especially for embedded-MLA systems, this step may need to be performed without the main lens. A point light source such as an LED is placed at effective infinity relative to the MLA. An image of the source is formed within each sub-image. The MLA's position and orientation are then adjusted until the point source appears as sharp as possible within sub-images across the entire array. The formation of point-like images for a point source located at infinity signifies that the mutual distance between MLA and sensor is now f_l. Notably, in many light-field cameras the in-plane rotation of the MLA (exaggerated in Fig. 2.21) is constrained but otherwise not explicitly adjusted.

Step 2 is used to establish the (s, t) coordinates of each microlens' centre across the entire image, such that data extraction could be performed at desired (u, v, s, t). This step quantifies the (s, t) positions and rotational orientation of the MLA relative to the sensor. Since the size and position of sub-images are actually not exactly identical

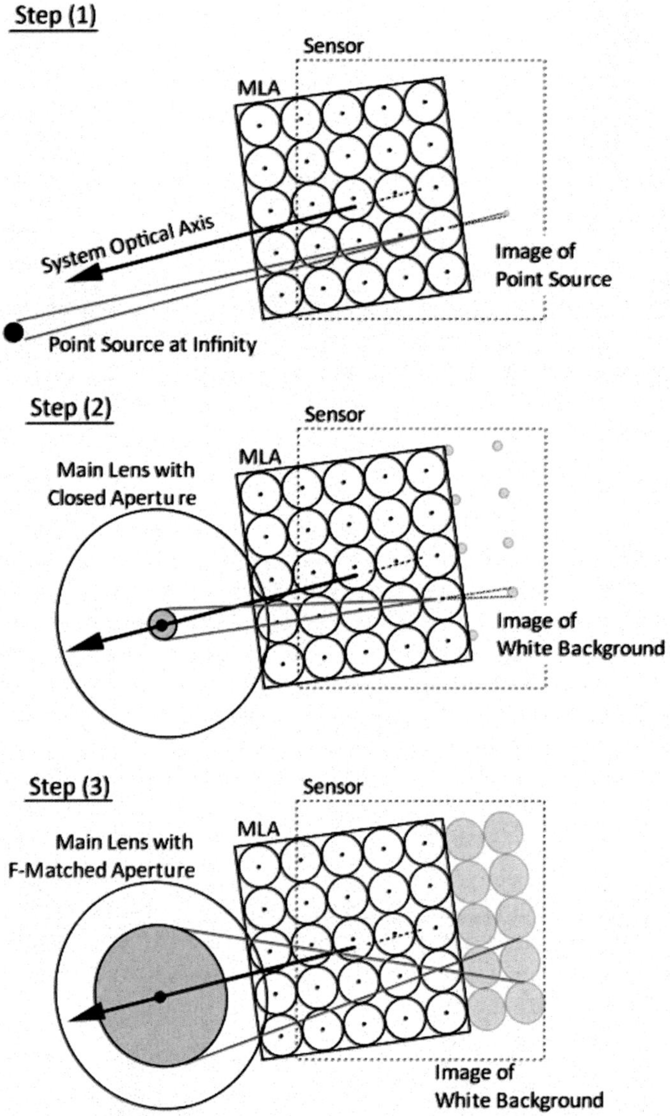

Fig. 2.21 Steps of setting up a light-field camera for imaging

to the physical microlens, but larger by a degree dependent on $S_i/d_{i,l} \approx d_{o,l}/d_{i,l}$, this step must be performed with the main lens installed and focused on the intended target to fix S_i. Next, a flat white subject (e.g. well-lit paper) is placed in front of the camera, and the main lens' aperture is turned down to a minimum. Since the size of each sub-image is correspondent to (u, v)-range, a minimised aperture creates a single point per microlens (for a flat white subject). Each point represents the

(s, t) position of the microlens centre on the sensor. An example of this image was shown in Fig. 2.19a. A straightforward image processing algorithm can then locate all bright points in the raw image and associate them with the MLA's microlens centre positions. This data is imperative for later decoding of light-field images into perspective views or refocused views.

Finally, step 3 is the F-matching procedures as alluded to in earlier sections. Without changing the main lens' focus, the aperture is widened again until adjacent sub-images contact but do not overlap. The light-field camera is now considered ready for data acquisition. Notably, for systems such as the DragonEye (Fig. 2.18a) with external adjustment screws, steps 1 and 2 may be performed together. Additionally, microlens centre-finding may also be performed based on the F-match image, if a circle-finding algorithm is instead employed.

Summary

The fundamental principles of light-field camera were introduced in this chapter using the thin lens equation and conventional imaging as a foundation for discussion. Concepts such as object and image spaces, nominal focal plane and depth-of-field were highlighted as being directly relevant to image formation for a light-field camera. The discussion then concentrated on unfocused light-field systems at length, given its prevalence in flow diagnostics, and key concepts such as (s, t) and (u, v) sampling, perspective view formation and refocused image formation were introduced. The focused light-field system was presented, but largely in reference to works in the wider computational imaging literature due to their still nascent application in flow diagnostics. Finally, synthetic image generation was introduced as an empirical approach to analyse and optimise camera design parameters. The hardware design for various light-field camera architectures was then discussed, followed by a procedural description on how to set up and calibrate an MLA within the hardware.

As of present, camera hardware remains split into the two general categories of embedded-MLA and relayed-MLA cameras, with the latter having lower performance due to aberrations and vignette introduced by relay lenses. However, the latter is still deemed the norm for pairing an MLA with high-speed cameras and critical for developing light-field intensified imagers. High-speed light-field cameras remain constrained by the pixel-to-speed trade-off for the foreseeable future, though progressively more powerful new sensors such as the Gpixel's Gsprint family of sensors with 5120×4096pixel in resolution and 1000FPS in speed are continuously expanding the design space for light-field devices.

References

Bolan J, Hall E, Clifford C, Thurow B (2016) Light-field imaging toolkit. SoftwareX 5:101–106

Butler R (2012) Lytro light field camera. Digit Photogr Rev. https://www.dpreview.com/reviews/lytro

Cho D, Lee M, Kim S, Tai Y-W (2013) Modeling the calibration pipeline of the lytro camera for high quality light-field image reconstruction. In: Proceedings of the IEEE international conference on computer vision. pp 3280–3287

Dansereau DG, Pizarro O, Williams SB (2013) Decoding, calibration and rectification for lenselet-based plenoptic cameras. In: Proceedings of the IEEE conference on computer vision and pattern recognition. pp 1027–1034

Dansereau DG, Schuster G, Ford J, Wetzstein G (2017) A wide-field-of-view monocentric light field camera. In: Proceedings of the IEEE conference on computer vision and pattern recognition. pp 5048–5057

Deem EA, Zhang Y, Cattafesta LN et al (2016) On the resolution of plenoptic PIV. Meas Sci Technol 27:84003

Drazic V, Sacré J-J, Schubert A et al (2012) Optimal design and critical analysis of a high-resolution video plenoptic demonstrator. J Electron Imaging 21:11007

Fahringer TW, Lynch KP, Thurow BS (2015) Volumetric particle image velocimetry with a single plenoptic camera. Meas Sci Technol. https://doi.org/10.1088/0957-0233/26/11/115201

Fahringer TW, Thurow BS (2012) Tomographic reconstruction of a 3-D flow field using a plenoptic camera. In: 42nd AIAA fluid dynamics conference and exhibit

Fahringer TW, Thurow BS (2018) Plenoptic particle image velocimetry with multiple plenoptic cameras. Meas Sci Technol. https://doi.org/10.1088/1361-6501/aabe1d

Georgiev T, Lumsdaine A (2010a) Reducing plenoptic camera artifacts. In: Computer graphics forum. Wiley Online Library, pp 1955–1968

Georgiev TG, Lumsdaine A (2010b) Focused plenoptic camera and rendering. J Electron Imaging 19:21106

Hall EM, Fahringer TW, Guildenbecher DR, Thurow BS (2018) Volumetric calibration of a plenoptic camera. Appl Opt 57:914–923

Liu Y, Hossain M, Sun J, et al (2017) Design a cage-typed light field camera system for flame measurement. IEEE Sens

Lumsdaine A, Georgiev T (2009) The focused plenoptic camera. In: 2009 IEEE international conference on computational photography (ICCP). IEEE, pp 1–8

Lynch K (2011) Development of a 3-D fluid velocimetry technique based on light field imaging

Ma H, Qian Z, Mu T, Shi S (2019) Fast and accurate 3D measurement based on light-field camera and deep learning. Sensors 19:4399

Markus (2012) Lytro specifications: a deeper look inside. Light Forum Light F Comput Imaging News. http://lightfield-forum.com/2012/11/lytro-specifications-a-deeper-look-inside/

Ng R (2006) Digital light field photography. Stanford Univ

O'brien S, Trumpf J, Ila V, Mahony R (2018) Calibrating light-field cameras using plenoptic disc features. In: 2018 International conference on 3D vision (3DV). pp 286–294

Pierce D (2014) Lytro illum review: this is the camera of the future. In: The verge. http://lightfield-forum.com/2012/11/lytro-specifications-a-deeper-look-inside/

Qi Q, Hossain MM, Zhang B et al (2019) Flame temperature reconstruction through a multi-plenoptic camera technique. Meas Sci Technol 30:124002

Schuster GM, Dansereau DG, Wetzstein G, Ford JE (2019) Panoramic single-aperture multi-sensor light field camera. Opt Express 27:37257–37273

Shi S, Ding J, New TH et al (2019) Volumetric calibration enhancements for single-camera light-field PIV. Exp Fluids. https://doi.org/10.1007/s00348-018-2670-5

Shi S, Wang J, Ding J et al (2016) Parametric study on light field volumetric particle image velocimetry. Flow Meas Instrum 49:70–88

Song X, Gu M, Cao L et al (2019) A microparticle image velocimetry based on light field imaging. IEEE Sens J. https://doi.org/10.1109/JSEN.2019.2927414

Tan ZP, Alarcon R, Allen J et al (2020) Development of a high-speed plenoptic imaging system and its application to marine biology PIV. Meas Sci Technol. https://doi.org/10.1088/1361-6501/ab553c

Tan ZP, Johnson K, Clifford C, Thurow BS (2019) Development of a modular, high-speed plenoptic-camera for 3D flow-measurement. Opt Express. https://doi.org/10.1364/oe.27.013400

Zhao Y, Li H, Mei D, Shi S (2020) Metric calibration of unfocused plenoptic cameras for three-dimensional shape measurement. Opt Eng. https://doi.org/10.1117/1.oe.59.7.073104

Zhao Z, Shi S (2021) Volumetric calibration for scheimpflug light-field PIV. Exp Fluids 62:1–18. https://doi.org/10.1007/S00348-021-03350-0

Chapter 3
Volumetric Calibration for Light-field Camera with Regular and Scheimpflug Lens

Shengxian Shi, Zhou Zhao, and T. H. New

Abstract This chapter describes the volumetric calibration methods for light-field camera with both regular and tilted main lens. By utilising such light-field image characteristics as plenoptic disc feature and point-like feature, methods that could accurately establish the correspondence between a tracer particle and its affected pixels regardless optical distortion will be detailed here. It will demonstrate how to calculate such calibration matrix with a dot array board, and how affected pixels of a tracer particle as well as its weighting coefficient could be more accurately determined with the calibration matrix, which will lay a solid foundation for 3D light-field particle image reconstruction.

Keywords Light-field calibration · Plenoptic disk feature · Scheimpflug light-field imaging · Scheimpflug calibration · Light-field distortion

Calibration for Light-field Camera with Regular Lens

For volumetric PIV, a core step is retrieveing three-dimensional information of seeding particles, which is normally achieved by the MART method. To accurately reconstruct a 3D particle image, a prerequisite is to correctly establish the mapping between seeding particle and its affected pixels. For Tomo-PIV, such mapping is a one-to-one relation and can be relatively easy determined by traditional camera calibration method (Wieneke 2008). For light-field PIV however, such mapping is complicated by the "optical coding" of microlens. As illustrated in Chap. 2, seeding

S. Shi · Z. Zhao (✉)
School of Mechanical Engineering, Shanghai Jiao Tong University, Shanghai 200240, China
e-mail: zhaozhou1994@sjtu.edu.cn

S. Shi
e-mail: kirinshi@sjtu.edu.cn

T. H. New
School of Mechanical and Aerospace Engineering, Nanyang Technological University, Singapore 639798, Singapore
e-mail: DTHNEW@ntu.edu.sg

S. Shi and T. H. New (eds.), *Development and Application of Light-Field Cameras in Fluid Measurements*, https://doi.org/10.1007/978-3-031-01779-7_3

particle and its affected pixels is a one-to-multiple mapping which varies with the particle's spatial location. If we inspect the light rays of a seeding particle closely by Gaussian optics (Fig. 3.1a), we could find such rays will form a plenoptic disk feature (here we follow the term defined by (O'brien et al. 2018) on the MLA plane (Fig. 3.1b), and eventually produce point-like features (the affected pixels) on sensor plane (Fig. 3.1c, d).

Follow the definition of MART (Eq. 3.1), reconstruct a 3D particle image means to iteratively calculate a 3D matrix $E(X_j, Y_j, Z_j)$ from the 2D matrix $I(x_i, y_i)$ (raw light-field particle image) with the help of weighting coefficient $w_{i,j}$. Weighting coefficient is the parameter that quantifies the mapping between a voxel (X_j, Y_j, Z_j) and its multiple affected pixels $(x_i, y_i)(i \in N_j)$. As illustrated in Fig. 3.1, accurate calculation of weighting coefficient $w_{i,j}$ relies on correctly identifying the affected pixels (point-like features in Fig. 3.1d) for a seeding particle P, which in turn requires precise determination of plenoptic disc feature (C_{df}, D_{df}), as well as the centre p_c of each point-like feature (Fig. 3.1b–d). In the next section, it will be detailed how to accurately calculate these parameters for each voxel even if optical distortions are

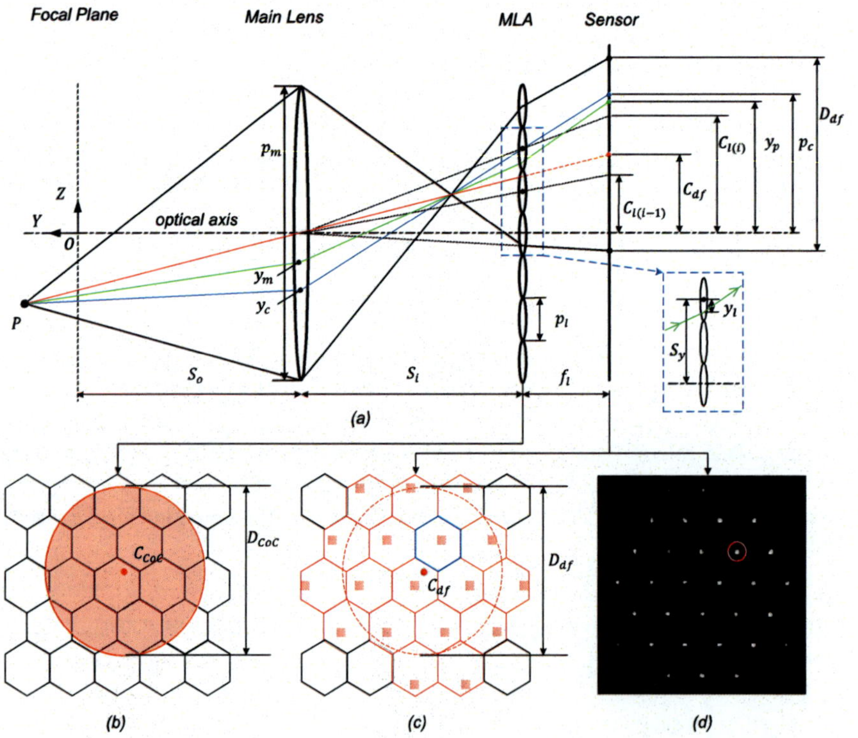

Fig. 3.1 **a** Ray propagation inside a light-field camera (only one-dimensional schematic is shown here); **b** plenoptic disc feature formed on the MLA plane (shown for a hexagonal lenslet); **c** plenoptic disc feature formed on the sensor plane; **d** synthetic light-field image showing distinctive point-like features

present.

$$E\left(X_j, Y_j, Z_j\right)^{k+1} = E\left(X_j, Y_j, Z_j\right)^k \left(\frac{I(x_i, y_i)}{\sum_{j \in N_i} w_{i,j} E\left(X_j, Y_j, Z_j\right)^k} \right)^{\mu w_{i,j}} \tag{3.1}$$

Determination of Plenoptic Disc Feature Diameter and Centre

If MLA is perfectly parallel to image sensor and the main lens is distortion free, we could then use thin-lens model to derive the diameter (D_{CoC}) of the plenoptic disc feature formed on the MLA plane (which can be called as the diameter of the circle of confusion, D_{CoC}) and the diameter (D_{df}) of plenoptic disc feature formed on the sensor plane (Eq. 3.2 and 3.3). Please refer to Fig. 3.1 and nomenclature for specific meaning of the symbols.

$$D_{\mathrm{CoC}} = p_m \left(\frac{S_i}{S_o} - \frac{S_i}{z + S_o} \right) = p_m \left(-M - \frac{1}{\frac{z}{S_i} - \frac{1}{M}} \right) \tag{3.2}$$

$$D_{df} = \frac{S_i + f_l}{S_i} \cdot p_m \left(-M - \frac{1}{\frac{z}{S_i} - \frac{1}{M}} \right) \tag{3.3}$$

For a perfect thin lens, we have:

$$S_i = f_m(1 - M) \tag{3.4}$$

$$f_\# = f_m / p_m \tag{3.5}$$

For a practical light-field camera, offset and/or relative rotation are very likely to present between MLA and image sensor. Further, light-field PIV experiments normally use fixed focal length main lens, which consists of a number of individual lenses. A thin-lens model may result in noticeable ray prediction error. To compensate ray calculation errors due to such optics distortion, five correction parameters (k_1, k_2, k_3, k_4, k_5) are introduced to modify Eq. 3.3. Consider the fact that the focal length of lenslet f_l is negligible when compared to the image distance S_i, the diameter of plenoptic disc feature D_{df} can be calculated according to the following equation:

$$D_{df} = k_1 \cdot p_m \left(-M - \frac{1}{\frac{k_2 z + k_3 x + k_4 y + k_5}{S_i} - \frac{1}{M}} \right) \tag{3.6}$$

As for the plenoptic disc feature centre C_{df}, it is defined as the intersection point that main light ray made with the sensor plane (e.g. the red light ray in Fig. 3.1a). According to Gaussian optics, propagation of a light ray (e.g. the green light ray in Fig. 3.1a) and where it hits the sensor plan can be calculated according to the following equation:

$$y_p = S_y + \frac{y_l + S_y - y_m}{S_i} f_l = C_{l(i)} - \frac{f_l}{S_i} y_m + \frac{y_l}{S_i} f_l \tag{3.7}$$

Using triangle similarity, the centre of plenoptic disc feature C_{df}, and the point-like feature centre, p_c, can be calculated from the following equations:

$$p_c = C_{l(i)} - \frac{f_l}{S_i} y_c \tag{3.8}$$

$$C_{df} = \frac{y_c}{p_m} D_{df} + C_{l(i)} \tag{3.9}$$

$$\frac{p_m}{S_i} = \frac{p_{l\_sensor}}{f_l} \tag{3.10}$$

Collecting these equations, the diameter and centre of plenoptic disc feature, centre of each point-like feature can be related with lenslet centre and lenslet pitch (in sensor plane) according to the following equation:

$$\left(C_{l(i)} - p_c\right) D_{df} + p_{l\_sensor}\left(C_{l(i)} - C_{df}\right) = 0 \tag{3.11}$$

To implement Eq. 3.6 and 3.11, a calibration board with white dot array is normally translated along light-field camera optical axis so as to cover the entire measurement volume. By taking light-field images for the calibration board at each discrete position, a set of spatial coordinates $(P(X, Y, Z))$, which are known precisely from the calibration board, and their corresponding point-like feature images would be obtained. For each calibration dot at a given translation position, there will be more than one lenslet that are fully covered by the plenoptic disc feature. As an example, there are five lenslets fully covered by the plenoptic disc feature in Fig. 3.1c, each of such lenslets produces a group of point-like features in the sensor plane (Fig. 3.1d). Therefore, there will be multiple equations formulated for Eq. 3.11 for each calibration dot. To solve D_{df} and C_{df} from these equations, we need to firstly calculate the lenslet centre $C_{l(i)}$, lenslet pitch (in sensor plane) p_{l_sensor} and centre of each point-like feature p_c. As detailed in (Ng 2006), lenslet centre can be determined by stepping down the main lens aperture to its minimum and capture a light-field image for a white plate. The image captured would be an array of white dots in a black ground and the lenslet centre coordinate $C_{l(i)}$ is then calculated by Gaussian fitting the centre of these white dots. On the other hand, centre of each point-like feature p_c can be calculated by Gaussian fitting with sub-pixel accuracy. Eventually, there will be multiple over-determined equations (e.g. five in Fig. 3.1c) can be used to solve D_{df} and C_{df} for every calibration dot $P(X, Y, Z)$.

With the calibration dot coordinates $P(X, Y, Z)$ and their corresponding plenoptic disc feature data (D_{df}, C_{df}), Eq. 3.6 can be solved for the calibration coefficients $(k_1, k_2, k_3, k_4, k_5)$ through the sequence quadratic program (SQP) method. Meanwhile, following equation can be solved for the mapping matrix $\mathcal{M}_{4\times3}$. Therefore, we could use these calibration coefficients and mapping matrix to determine plenoptic disc feature diameter D_{df} and centre coordinate C_{df} for any points in the measurement volume, which will be applied in the following section to calculate weighting coefficient $w_{i,j}$.

$$\begin{pmatrix} C_{df}(x) \\ C_{df}(y) \\ 1 \end{pmatrix} = \mathcal{M}_{4\times3} \begin{pmatrix} X \\ Y \\ Z \\ 1 \end{pmatrix} \tag{3.12}$$

Calculation of the Weighting Coefficient

Light rays from a seeding particle or a calibration dot would firstly be intercepted by MLA and then hit the sensor plane, naturally the weighting coefficient $w_{i,j}$ is divided into two portions—lenslet weighting coefficient w_1 and pixel weighting coefficient w_2 ($w_{i,j} = w_1 \times w_2$). w_1 is the overlap area between the affected lenslet and plenoptic disc feature (i.e. overlap area between red dotted circle and lenslets in Fig. 3.1c), and w_2 is the overlap area between the affected pixels and point-like features (i.e. white dots in Fig. 3.1d).

The previous step has calculated lenslet centre $C_{l(i)}$, plenoptic disc feature centre C_{df} and diameter D_{df}, from which lenslets affected by a seeding particle (or a voxel $E(X_j, Y_j, Z_j)$ to be precisely) can be correctly identified (i.e. the red lenslets in Fig. 3.1c). The weighting coefficient w_1 of each lenslet could be determined by directly calculating its overlap area with the plenoptic disc feature (e.g. the red dotted circle in Fig. 3.1c), or using the Monte Carlo method to ease the code implementation. In such method, certain number of random points are generated inside the plenoptic disc feature, and w_1 is calculated as the percentage of points that fall into the specific lenslet. The choice of random points number is a balance between accuracy and computational cost.

Calculating the pixel weighting coefficient w_2 fundamentally requires the affected pixels be precisely located for each point-like feature in Fig. 3.1d. Let's take one point-like feature (e.g. the one highlighted in red in Fig. 3.1d) as an example to illustrate the detailed process. Trace back to its corresponding lenslet (e.g. the one highlighted in blue in Fig. 3.1c), the affected pixels should fall in the range covered by light rays passing through its upper and lower boundary ($-p_l/2 \leq y_l \leq p_l/2$, see

Fig. 3.1a, demonstrate in one-dimension), and such range can be calculated according to Eq. 3.7. Note that the image distance is far larger than lenslet focal length and lenslet pitch ($S_i \gg f_l$, $S_i \gg p_l$), therefore, $\frac{y_l}{S_i} f_l$ in Eq. 3.7 can be safely neglected. Take a typical light-field PIV system as an example, the focal length of main lens and lenslet are around $f_m = 200$ mm, $f_l = 0.3$ mm, pitch of lenslet and pixel are around $p_l = 0.08$ mm, $p_p = 5.5\ \mu m$. Thereby $\frac{y_l}{S_i} f_l < \frac{0.08}{2 \times 200} \times 0.3$ mm $= 0.00006$ mm, which is less than 0.01pixel. Furthermore, the relation between y_l and plenoptic disc feature parameters can be established using triangle similarity.

$$\frac{y_m}{p_m} = \frac{C_{df} - \left(C_{l(i)} + y_l \frac{S_i + f_l}{S_i}\right)}{D_{df}} \tag{3.13}$$

Finally, the location where a light ray hits on the sensor plane y_p can be derived from Eqs. 3.7, 3.10 and 3.13.

$$y_p = C_{l(i)} + \frac{p_{l\_sensor}\left(y_l \frac{S_i + f_l}{S_i} + C_{li} - C_{df}\right)}{D_{df}} \tag{3.14}$$

If we substitute $y_l = y_{p\_dn} = -\frac{d_l}{2}$ and $y_l = y_{p\_up} = \frac{d_l}{2}$ to the above equation, the affected pixels should fall in the region defined by Eqs. 3.15 and 3.16. Likewise, the weighting coefficient w_2 could be determined by directly calculating the overlap area between each affected pixel and the region determined from Eqs. 3.15 and 3.16, or apply a similar Monte Carlo method for a more realisable coding. To implement, certain number of random points are generated inside the determined region, w_2 is then calculated as the percentage of points that fall into the specific pixel.

$$y_{p\_dn} = \begin{cases} C_{l(i)} + \dfrac{p_{l\_sensor}\left(-\frac{d_l}{2} + C_{l(i)} - C_{df}\right)}{D_{df}}, & z < 0 \\ C_{l(i)} + \dfrac{p_{l\_sensor}\left(\frac{d_l}{2} + C_{l(i)} - C_{df}\right)}{D_{df}}, & z > 0 \end{cases}. \tag{3.15}$$

$$y_{p\_up} = \begin{cases} C_{l(i)} + \dfrac{p_{l\_sensor}\left(\frac{d_l}{2} \cdot + C_{l(i)} - C_{df}\right)}{D_{df}}, & z < 0 \\ C_{l(i)} + \dfrac{p_{l\_sensor}\left(-\frac{d_l}{2} \cdot + C_{l(i)} - C_{df}\right)}{D_{df}}, & z > 0 \end{cases} \tag{3.16}$$

For an overview of the light-field volumetric calibration method, the above procedures can be summarised as a flowchart in Fig. 3.2. The dense ray tracing-based MART reconstruction method (DRT-MART) is included here to demonstrate a close loop. The DRT-MART method will be detailed in Chap. 4.

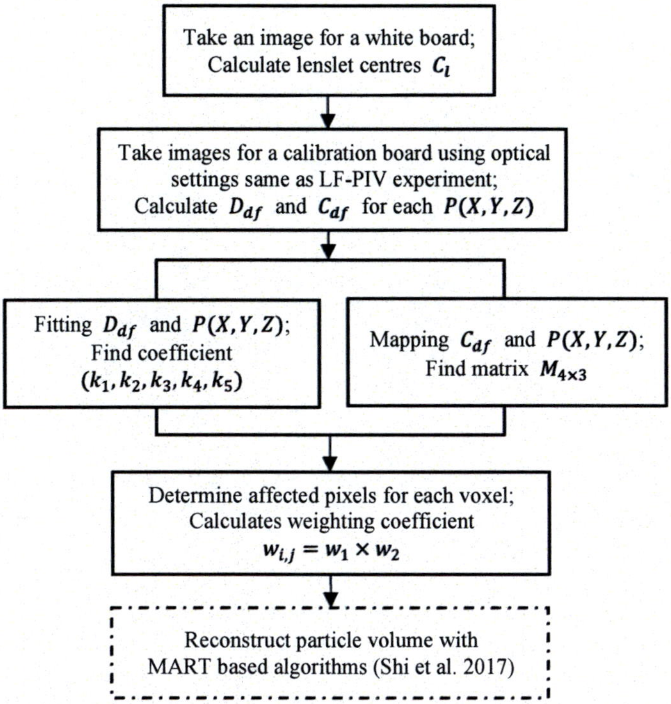

Fig. 3.2 Flowchart of the volumetric calibration method (steps in black boxes), the reconstruction step is also shown in dotted box for completeness. *Source* Shi et al. 2017

Simulation Studies

To demonstrate performance of the volumetric calibration method, a series of simulation tests were performed with synthetic light-field images, which are generated using distortion free, thin-lens model. The calibration method was validated by directly comparing synthetic images with the affected pixels which are predicted by the calibration model.

Optical parameters for generating synthetic light-field images are identical to the in-house 29 M light-field camera (Table 3.1). To be specific, an artificial calibration board measures 36.3×24.2mm and contains 13×9 white dots array (2.5mm grid spacing) was traversed along optical axis (z-axis) in a range of -10 mm $\leq z \leq$ 10 mm (0.5 mm step-interval). Follow procedures detailed in Chap. 2, synthetic light-field images were generated by propagating a large number of light rays (e.g. 5,000,000 light rays) from a calibration dot (simulated as a point light source), through the main lens and MLA, until they reach the pixel plane. In such way, 41 synthetic light-field calibration images were generated such that there are in total $13 \times 9 \times$

Table 3.1 Optical parameters for synthetic light-field image generation

Parameter	Value
Pixel resolution	6600 × 4400
Pixel size	5.5 μm
MLA resolution	450 × 350
Lenslet size	77 μm
Lenslet focal length	308 μm
Main lens focal length	200 mm
Main lens f number	4
Simulation volume	36.3 × 24.2 × 20 mm
Magnification	− 1

41 = 4797 calibration dots available to calculate coefficients $(k_1, k_2, k_3, k_4, k_5)$ and mapping matrix $\mathcal{M}_{4\times3}$ from Eqs. 3.6 and 3.12. Follow procedures in Fig. 3.2, the calibration coefficient and mapping matrix were determined as below for the above synthetic light-field image set.

$$(k_1, k_2, k_3, k_4, k_5) = (1.009, 1.0021, -0.0007, 0.0016, 0.0086)$$

$$\mathcal{M}_{4\times3} = \begin{bmatrix} -1.5467e+002 & +1.1081e-004 & +7.6924e+000 & +3.2995e+003 \\ -1.6632e-003 & +1.5467e+002 & +5.0254e+000 & +2.1995e+003 \\ -9.0152e-007 & +1.4232e-007 & +2.3104e-003 & +1.0000e+000 \end{bmatrix}$$

We could take one calibration dot (e.g. $P(X, Y, Z) = (0, 0, -2.0)$mm) to validate the calibration results. The calibration dot synthetic light-field image is shown in Fig. 3.3a, its calculated weighting coefficient w is plotted as a grey scale image (normalise and convert into 0–255) in Fig. 3.3b. A side-by-side comparison reveals that the calculated weighting coefficient faithfully reflects the location and extent of affected pixels for a calibration dot. For a more quantitative evaluation, the affected

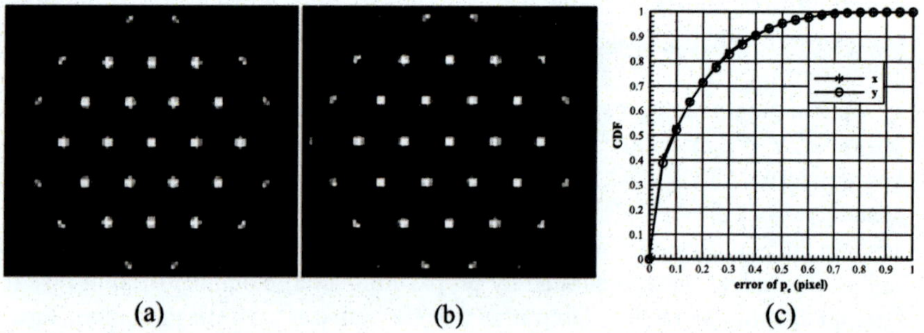

Fig. 3.3 **a** Synthetic light-field image of a calibration dot at $(P(X, Y, Z) = (0.0 - 2.0)$mm); **b** corresponding weighting coefficient calculated with the proposed calibration method (plotted as grey scale image for better comparison); **c** error distribution of the calculated point-like feature centre p_c

pixel clusters were identified by the calibration method for each calibration dot, and their centre coordinates p_c are calculated using weighted centroid method. On the other hand, centre coordinates of point-like features are directly calculated from every calibration dot synthetic light-field image and use as the ground truth. The cumulative distribution function (CDF) of the absolute errors between p_c and the ground truth is calculated for all 41 synthetic light-field calibration images. As shown in Fig. 3.3c, over 95% of identified pixel clusters have an absolute error lower than 0.5 pixel, which proves a high calibration accuracy for the current method.

Calibration Experiment

Simulation studies could only validate the calibration method under ideal conditions, calibration experiments are carried out here to further test its performance when lens aberration and optical window distortions are present. A typical light-field camera calibration setup is shown in Fig. 3.4a, which consists of a light-field camera, a glass that is identical to optical window in the test section, a linear translation stage and a calibration board.

A 21×15 white calibration dot array (0.2 mm diameter) with a grid spacing of 2.5 mm was laser-etched on a matt black aluminium board (Fig. 3.4b). The calibration board was traversed by a Thorlabs LNR50S/M linear translation stage (resolution 0.1 μm) to-and-fro of the focal plane to cover the entire measurement area (e.g. $-10 - 10$ mm). Calibration images were taken at 0.5 mm step-interval with optical setting (e.g. main lens type, focal distance, main lens aperture) exactly same as light-field PIV experiments, such that any optical imperfections during the experiment are faithfully recorded in the calibration images. In total, there are 41 light-field calibration images captured by the in-house 29 M light-field camera (with Micro-NIKKOR 200 mm lens), which produces $21 \times 15 \times 41 = 12915$ calibration dots to solve Eqs. 3.6 and 3.12. The calculated calibration coefficient and mapping matrix for current experiment are:

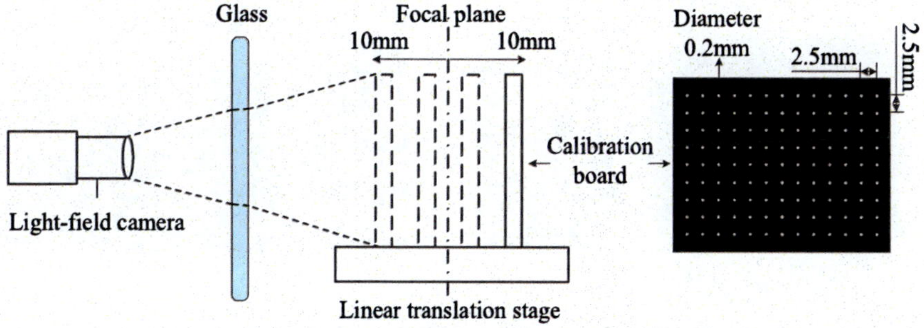

Fig. 3.4 **a** Schematic of the calibration setup, and **b** the calibration board

$$(k_1, k_2, k_3, k_4, k_5) = (0.7322, 0.9882, 0.0145, 0.0231, -0.2)$$

$$\mathcal{M}_{4\times3} = \begin{bmatrix} +1.7765e+002 & +1.2227e-001 & +4.9384e+000 & +3.3106e+003 \\ -3.7039e-001 & +1.7740e+002 & +1.6402e+000 & +2.2286e+003 \\ +6.5267e-006 & -2.3101e-005 & +8.2910e-004 & +1.0000e+000 \end{bmatrix}$$

With these calibration data, pixel clusters affected by each calibration dots can be determined by Eqs. 3.15 and 3.16, and centre p_c of each cluster can be calculated by weighting centroid method (e.g. green circle signs in Fig. 3.5). To evaluate the calibration performance, pixel cluster centre p_c is compared with two sets of data, one being calculated directly from light-field calibration images (ground truth) and the other one predicted by a ray tracing method (reference data). The ground truth data were calculated with weighted centroid method for each real bright pixel cluster (e.g. red plus signs in Fig. 3.5). For the reference data, affected pixel clusters for each calibration dot are identified by tracing 5,000,000 random light rays from the calibration dot spatial location to the sensor plane (with thin lens model, in a similar way as generating a synthetic light-field image). Again, the reference data are calculated from the estimated pixel clusters with weighted centroid method (e.g. yellow cross signs in Fig. 3.5). The ray tracing method does not take any optical imperfections into account. As Fig. 3.5 shows, the reference data deviate significantly from the ground truth, indicating optical distortions cannot be neglected. On the other hand, pixel cluster centres predicted by the calibration method match generally well with the ground truth.

For a more quantitative assessment, pixel cluster centres are calculated for all 41 light-field calibration images, and compared with the ground truth for both calibration and ray tracing methods. The CDF of p_c absolute error (in x- and y-direction) is plotted in Fig. 3.6a for both calibration and ray tracing methods, which demonstrates

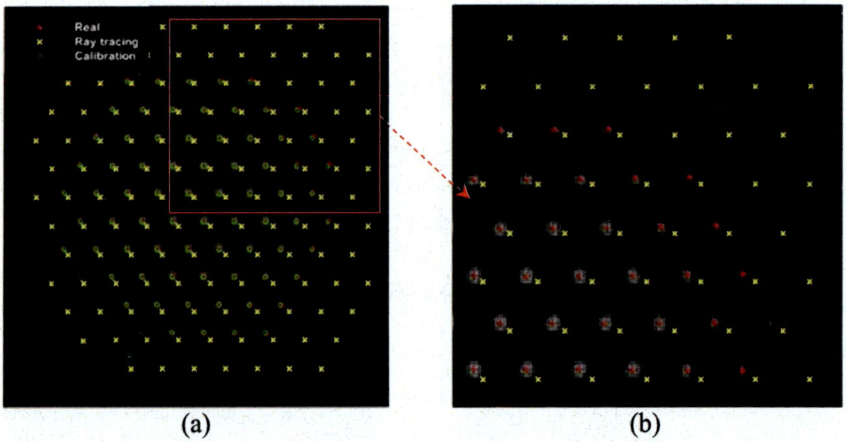

Fig. 3.5 **a** An example of light-field image for one calibration dot, showing the real centre of affected pixels, together with results from ray tracing and volumetric calibration method, **b** magnified view for a better appreciation of the differences between two methods

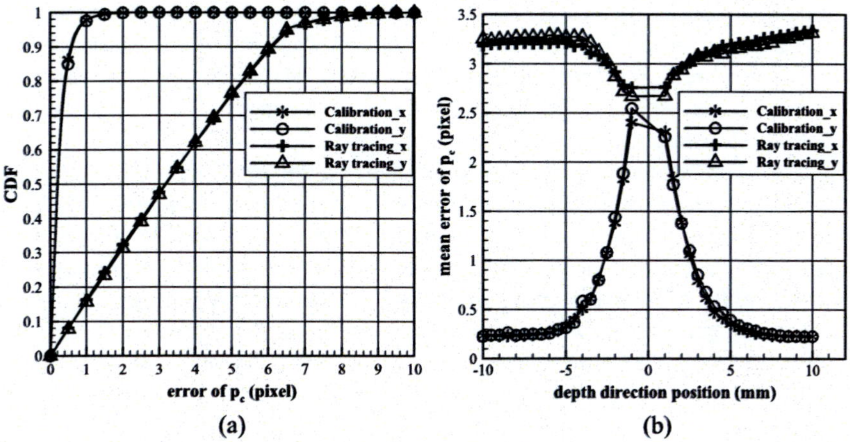

Fig. 3.6 **a** CDF error distribution of the detected point-like feature centre, **b** error distribution of the detected point-like feature centre along depth direction

that about 98% affected pixel clusters can be correctly identified (with absolute p_c error less than 1 pixel). Whereas, there are less than 18% affected pixel clusters can be correctly identified by the ray tracing method, due to the significant optical distortions in real experiments.

However, such data only represents calibration performance for discrete locations since calibration board images were only taken at 41 positions. To find out how accurate predictions can be achieved by the calibration coefficient $(k_1, k_2, k_3, k_4, k_5)$ and mapping matrix $\mathcal{M}_{4\times3}$ for entire measurement volume, the absolute p_c errors are plotted along optical axis for both calibration and ray tracing methods (Fig. 3.6b). Although the calibration method performs generally better than the ray tracing one, it cannot actually predict the affected pixel clusters for the entire measurement volume. Only in the regions $(-10 \le z \le -3)$ and $(3 \le z \le 10)$ where affected pixel clusters can be correctly identified (absolute p_c error less than 1 pixel), for areas near the focal plane, there is a distinct increase in p_c errors. Recall the principle of light-field imaging (Chap. 2), for calibration dots near the focal plane, light rays would focus on fewer lenslets, and the extreme scenario happens at the origin of focal plane $(O(0, 0, 0))$, where light rays will focus on only one lenslet. Consequently, there will be few (or only one at $O(0, 0, 0)$) lenslets fully covered by plenoptic disc feature, and hence fewer numbers of Eq. 3.11 are available to calculate D_{df} and C_{df}.

Calibration for Light-field Camera with Scheimpflug Lens

There are situations where Scheimpflug lens is necessary for light-field imaging, e.g. off-axis imaging due to limited optical access, and to align the focal plane for dual light-field camera configuration. Scheimpflug lens will further complicate ray propagation inside a light-field camera and significantly alter the mapping relation between a seeding particle to its affected pixels. We shall again use the unique plenoptic disk feature and point-like features, and jointly apply Scheimpflug imaging principle to establish the mapping relationship between a voxel and its affected pixels.

Based on thin lens model and Scheimpflug imaging principle (Merklinger 1996), the optical model of a Scheimpflug light-field camera is illustrated in Fig. 3.7. As it shows, main lens plane, focal plane and MLA plane intersect along one line (the Scheimpflug line, L_s), which is determined by Scheimpflug rule. In addition, parallel-to-film plane (PTF plane) and focal plane intersect along the Hinge line, and the separation between main lens plane and Hinge line is f_m (Hinge rule). Therefore, once the tilted angle (γ) between main lens plane and MLA plane is set, a Scheimpflug light-field configuration is uniquely determined.

Refer to Fig. 3.7a, the coordinate frame of Scheimpflug light-field camera is defined as follows, $O-XYZ$ denotes lens coordinate system (LCS), point $O(0, 0, 0)$ is main lens optical centre, and OZ is the optical axis. $O_m(0, 0, -S_i)$ is the intersection point between optical axis and MLA plane. In addition, we define $O_p - UV$ as 2D pixel coordinate system (PCS), where point O_p is the projection of O_m on sensor plane. To simplify the discussion, MLA coordinate system will not be independently defined, instead it will be indirectly represented in PCS. For instance, the coordinate of O_m will be described as $O_m\left(O_p^u, O_p^v\right)$ in the following part.

Based on such coordinate systems, we use (P^x, P^y, P^z) to denote the coordinate of a seeding particle P, and $Q(Q^x, Q^y, Q^z)$ to represent coordinate for its converging imaging point. From thin lens model, we have

$$\frac{1}{S_o'} + \frac{1}{S_i'} = \frac{1}{P^z} + \frac{1}{-Q^z} = \frac{1}{f_m} \tag{3.17}$$

where P^z is the Z-direction distance from seeding particle P to main lens, $-Q^z(S_i')$ is the Z-direction distance from Q to main lens, and f_m is the focal length of main lens. Also, there always exists a light ray (the blue dashed line POQ in Fig. 3.7a) passing through seeding particle P, main lens centre O, and converging point Q. From triangle similarity and Eq. 3.17, the relationship between P and Q can also be modelled as

$$\begin{bmatrix} Q^x \\ Q^y \\ Q^z \end{bmatrix} = \frac{-S_i'}{S_o'} \begin{bmatrix} P^x \\ P^y \\ P^z \end{bmatrix} = \frac{f_m}{f_m - P^z} \begin{bmatrix} P^x \\ P^y \\ P^z \end{bmatrix} \tag{3.18}$$

Recall that the angle between main lens plane and MLA plane is γ, the projection of Q on MLA plane $Q_1(Q^u, Q^v)$ can be represented as:

$$\begin{bmatrix} Q^u \\ Q^v \end{bmatrix} = \begin{bmatrix} O_p^u + \frac{1}{p_p} Q^x \\ O_p^v - \frac{1}{p_p}\left[(S_i + Q^z)\sin\gamma - Q^y \cos\gamma\right] \end{bmatrix} \tag{3.19}$$

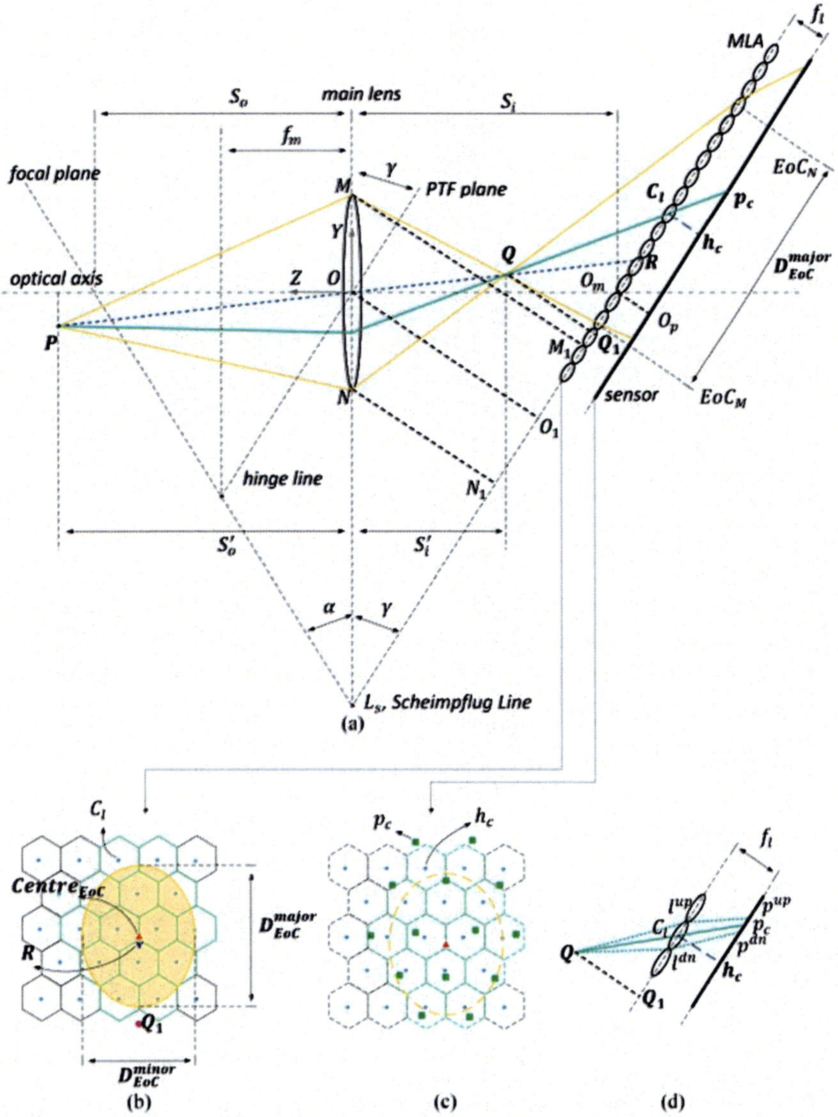

Fig. 3.7 **a** Optical configuration and light propagation process of the Scheimpflug light-field camera (only one-dimensional schematic is shown here), **b** plenoptic disc feature corresponding to a point *P* formed on the MLA plane, **c** point-like features corresponding to a point *P* formed on the sensor plane, **d** the region of the point-like pattern can be determined by the boundaries of the micro-lens

The distance Q^d between Q and Q_1 can be modelled as:

$$\begin{aligned} Q^d &= \frac{1}{p_p}\left[Q^y \sin\gamma + (S_i + Q^z)\cos\gamma\right] \\ &= \frac{1}{p_p}\left[\frac{f_m}{f_m - P^z}P^y \sin\gamma + \left(S_i + \frac{f_m}{f_m - P^z}P^z\right)\cos\gamma\right] \end{aligned} \tag{3.20}$$

Due to the inclination of MLA and sensor plane, plenoptic disc feature formed on MLA plane is elliptic instead of circular (Fig. 3.7b, (Evens 2008a, b)), and the affected pixels (point-like features) are scattered in an elliptic region (Fig. 3.7c). From Fig. 3.7a, coordinate of point-like feature p_c can be derived with similar triangles (Eq. 3.21), where lenslet centre C_l can be determined in a similar fashion as previous section. Notice that the calculated coordinate of C_l is actually its perpendicular projection (h_c^u, h_c^v) on the sensor plane.

$$\begin{bmatrix} p_c^u \\ p_c^v \end{bmatrix} = \begin{bmatrix} h_c^u \\ h_c^v \end{bmatrix} + \frac{f_l}{Q_d}\left(\begin{bmatrix} h_c^u \\ h_c^v \end{bmatrix} - \begin{bmatrix} Q^u \\ Q^v \end{bmatrix}\right) \tag{3.21}$$

To further identified the affected pixels (point-like features), it will firstly need to determine the region that covered by affected lenslet, which could be achieved by tracing light rays that pass the edge of lenslet. Take the lenslet shown in Fig. 3.7d as an example, its edge can be defined as $l^{up}\left(h_c^u, h_c^v + \frac{p_l}{2p_p}\right)$ and $l^{dn}\left(h_c^u, h_c^v - \frac{p_l}{2p_p}\right)$. According to Gaussian optics, light ray $\rightarrow_{Ql^{up}}$ (the upper green dashed line in Fig. 3.7d) can be denoted as $\begin{pmatrix} h_c^u \\ h_c^v + \frac{p_l}{2p_p} \\ \mu_{up} \\ \sigma_{up} \end{pmatrix}$, where $\begin{pmatrix} \mu_{up} \\ \sigma_{up} \end{pmatrix}$ is the angle that the light ray made with MLA plane. Location where this light ray hits on sensor plane can be calculated according to:

$$\begin{pmatrix} p^{up,u} \\ p^{up,v} \\ \mu'_{up} \\ \sigma'_{up} \end{pmatrix} = \begin{pmatrix} 1 & & f_l & \\ & 1 & & f_l \\ & & 1 & \\ & & & 1 \end{pmatrix}\begin{pmatrix} 1 & & & \\ & 1 & & \\ -1/f_l & & 1 & \\ & -1/f_l & & 1 \end{pmatrix}\begin{pmatrix} h_c^u \\ h_c^v + \frac{p_l}{2p_p} \\ \mu_{up} \\ \sigma_{up} \end{pmatrix} + \begin{pmatrix} h_c^u \\ h_c^v \\ 0 \\ 0 \end{pmatrix} \tag{3.22}$$

and $\begin{pmatrix} \mu_{up} \\ \sigma_{up} \end{pmatrix}$ can be calculated as

$$\begin{pmatrix} \mu_{up} \\ \sigma_{up} \end{pmatrix} = \begin{pmatrix} \frac{h_c^u - Q^u}{Q^d} \\ \frac{h_c^v + \frac{p_l}{2p_p} - Q^v}{Q^d} \end{pmatrix} \tag{3.23}$$

Therefore, region of point-like features (affected pixels) that confined between light rays $\overrightarrow{Ql^{up}}$ and $\overrightarrow{Ql^{dn}}$ can be calculated from following equations:

$$\begin{cases} p^{up} = \begin{pmatrix} h_c^u \\ h_c^v \end{pmatrix} + f_l \begin{pmatrix} \frac{h_c^u - Q^u}{Q^d} \\ \frac{h_c^v + \frac{p_l}{2p_p} - Q^v}{Q^d} \end{pmatrix} \\ p^{dn} = \begin{pmatrix} h_{ci}^u \\ h_{ci}^v \end{pmatrix} + f_l \begin{pmatrix} \frac{h_{ci}^u - Q^u}{Q^d} \\ \frac{h_{ci}^v - \frac{p_l}{2p_p} - Q^v}{Q^d} \end{pmatrix} \end{cases} \tag{3.24}$$

Plenoptic Disk Feature Extraction

Let's now trace back to MLA plane to find out affected lenslets, which are covered by an elliptic plenoptic disk feature (Fig. 3.7b). To do that, we trace light rays that pass through edge $\left(M\left(0, \frac{p_m}{2}, 0\right), N\left(0, -\frac{p_m}{2}, 0\right)\right)$ of the main lens (Fig. 3.7a). Locations where they hit the MLA plan can be calculated from similar triangles. To differentiate from circular plenoptic disk feature, we rename the elliptic one as ellipse of confusion (EoC).

$$\begin{cases} \frac{QQ_1}{MM_1} = \frac{\mathrm{EoC}_M Q_1}{\mathrm{EoC}_M M_1} \\ \frac{QQ_1}{NN_1} = \frac{\mathrm{EoC}_N Q_1}{\mathrm{EoC}_N N_1} \end{cases} \tag{3.25}$$

The centre of EoC can be formulated as

$$C_{\mathrm{EoC}} = \frac{1}{2}(\mathrm{EoC}_M + \mathrm{EoC}_N) \tag{3.26}$$

The length of EoC major axis be calculated from

$$D_{\mathrm{EoC}}^{\mathrm{major}} = \frac{1}{2}|\mathrm{EoC}_M \mathrm{EoC}_N| \tag{3.27}$$

Likewise, if we trace light rays through the edge of main lens along X-axis $(G\left(\frac{p_m}{2}, 0, 0\right), H\left(-\frac{p_m}{2}, 0, 0\right)$, location where they reach the MLA plane (EoC_G, EoC_H) can be determined by similar triangles. Length of the EoC minor axis can be calculated as

$$D_{\mathrm{EoC}}^{\mathrm{minor}} = \frac{1}{2}|\mathrm{EoC}_G \mathrm{EoC}_H| \tag{3.28}$$

Calibration for Optical Distortions

At this point, the affected lenslets and pixels for any given seeding particle (spatial voxel) P can be determined from distortion free thin-lens models (Eqs. 3.24, 3.26–3.28). However, these models are rather ideal and cannot be applied directly. Firstly, the intersection point O_m between optical axis and MLA plane is normally difficult to determine. Secondly, tilted-angle marked on Scheimpflug adapter can only serve as a guidance, accurate angles γ_x (sensor rotation angle along X-axis) and γ_y (sensor rotation angle along Y-axis) are impossible to derive from the marked numbers. Finally, there are optical distortions caused by main lens, optical window as well as misalignment between MLA and sensor plane.

Figure 3.8 plots a modified optical model that takes main lens imperfection into account. If we set the main lens aperture to minimum, and take a light-field image for a white board, the recorded pattern would be the projection of light rays that pass through main lens aperture centre A and lenslet centre (e.g. g_c and the red dotted line in Fig. 3.8). Further, if we take a light-field image for a collimator without the Scheimpflug main lens (collimator-pattern image), the perpendicular projection h_c of micro-lens centres C_l can be calculated from such image. As such, the perpendicular projection $A_1(A^u, A^v)$ of aperture centre A and the distance $AA_1(A^d)$ can be determined by similar triangles.

$$\frac{g_c h_c}{l_c A_1} = \frac{f_l}{AA_1} \tag{3.29}$$

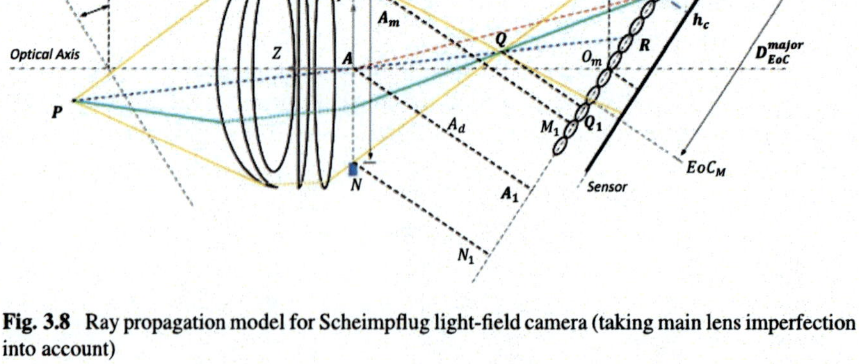

Fig. 3.8 Ray propagation model for Scheimpflug light-field camera (taking main lens imperfection into account)

From the collimator-pattern image, each lenslet can provide one equation for $A_1(A^u, A^v)$ and $AA_1(A^d)$, hence the aperture centre A can be accurately determined by solving the overdetermined Eq. 3.29 with least square method. Consequently, we can determine the intersection point $R(R^u, R^v)$ which is made by light ray $\rightarrow_{PAQ}$ (the blue dotted line in Fig. 3.8) and MLA plane. Studies have shown that difference between R and the C_{EoC} (Fig. 3.7b) is small and can be neglected (Evens 2008a, b).

Before deriving corrected model for plenoptic disk feature and point-like features, we should update main lens aperture p_m with effective aperture A_m, which is given by Eqs. 3.30 and 3.31.

$$\frac{A_m}{\frac{A^d}{\cos\gamma}} = \frac{p_l}{f_l} = \frac{1}{f_\#(1-M)} \tag{3.30}$$

$$A_m = \frac{A^d}{\cos\gamma}\frac{p_l}{f_l} \tag{3.31}$$

Since the tilted-angle cannot be precisely acquired, a correction factor k is introduced to compensate the prediction error of the major/minor axis length of EoC:

$$\begin{cases} D_{\text{EoC}}^{\text{major}} = k \times \frac{1}{2}|\text{EoC}_M\text{EoC}_N| \\ D_{\text{EoC}}^{\text{minor}} = k \times \frac{1}{2}|\text{EoC}_G\text{EoC}_H| \end{cases} \tag{3.32}$$

In the next, the pixel coordinate of $Q_1(Q^u, Q^v)$ can be derived as $\triangle\ QRQ_1 \sim\triangle\ ARA_1$

$$Q^{u,v} = R^{u,v} - \frac{Q^d}{A^d}R^{u,v} \tag{3.33}$$

Recall theoretical Eq. 3.20, Q^d is a function of $f\left(Q^u, Q^v, P^z, \gamma_x, \gamma_y, f_m, p_p\right)$. To implement, we use the three-order polynomial to approximate:

$$\begin{aligned} Q^d = {} & a_0 + a_1R^u + a_2R^v + a_3P^z + a_4R^uR^v + a_5R^uP^z + a_6\left(P^z\right)^2 \\ & + a_7R^vP^z + a_8(R^v)^2 + a_9\left(R^u\right)^2 + a_{10}\left(R^u\right)^2R^v \\ & + a_{11}(R^v)^2R^u + a_{12}\left(R^u\right)^2P^z + a_{13}(R^v)^2P^z \\ & + a_{14}\left(P^z\right)^2R^v + a_{15}\left(P^z\right)^2R^u + a_{16}R^uR^vP^z \\ & + a_{17}\left(R^u\right)^3 + a_{18}(R^v)^3 + a_{19}\left(P^z\right)^3 \end{aligned} \tag{3.34}$$

To calculate polynomial coefficients a_{0-19}, a calibration board is implied and calibration images are captured in a similar fashion as previous sections. Processing these images with Eqs. 3.21, 3.27, 3.28, 3.33, we can calculate $D_{\text{EoC}}^{\text{major}}$, $D_{\text{EoC}}^{\text{minor}}$, $Q\left(Q^u, Q^v, Q^d\right)$ and $R(R^u, R^v)$. As spatial points $P(P^x, P^y, P^z)$ are known precisely from the calibration board, we can solve Eq. 3.32 and Eq. 3.34 for correction factor k and polynomial coefficients a_{0-19}, respectively. Further, the data set $P(P^x, P^y, P^z)$ and $R(R^u, R^v)$ can be used to determine the mapping matrix $M_{4\times3}$ according to:

$$\begin{pmatrix} R^u \\ R^v \\ 1 \end{pmatrix} = M_{4\times3} \begin{pmatrix} P^x \\ P^y \\ P^z \\ 1 \end{pmatrix} \tag{3.35}$$

To summarise, the flow chat of calibration method is shown in Fig. 3.9, where the DRT-MART reconstruction step is included for the sake of completeness.

Calibration Experiment

To verify the calibration method, tests were performed with an experimental system shown in Fig. 3.10. The system consists of an in-house 29 M light-field camera, Micro-NIKKOR 200 mm main lens which was tilted to around 5° via a commercial Scheimpflug adaptor, a calibration board containing 21 × 15 white dot array (same as previous section). The calibration board was traversed to-and-fro of the focal plane in the range of −12.5–12.5 mm with a 0.5 mm step-interval along the Z_w-axis ((Thorlabs LNR50S/M, resolution 0.1 μm). There are in total 51 light-field calibration images were captured (21 × 15 × 51 = 16065 calibration points).

Following procedures described in Fig. 3.9, plenoptic disc feature centre $R(R^u, R^v)$ and diameter $\left(D_{\mathrm{EoC}}^{\mathrm{major}}, D_{\mathrm{EoC}}^{\mathrm{minor}}\right)$, as well as converging imaging point $Q\left(Q^u, Q^v, Q^d\right)$ are calculated for every calibration dot. Then the calibration coefficients a_{0-19}, correction factor k and mapping matrix $M_{4\times3}$ can be obtained. To evaluate the calibration performance, centre of every point-like feature in the calibration image is calculated with weighted centroid method, and is served as ground truth (e.g. green Δ in Fig. 3.11). On the other hand, centre of affected pixel cluster which is identified by the calibration method, is marked as red $*$ for a direct comparison. Figure 3.11 plots such data for a calibration dot at (−3, −3, −12.5), which demonstrates that, especially the zoom-in section, the affected pixel clusters can be correctly identified by the calibration method, and their predicted centre p_c matches well with ground truth.

For a more quantitative assessment, 51 calibration images were processed with the calibration method, and the predicted pixel cluster centres are compared with ground truth. The absolute errors in $u-$ and $v-$direction as well as nominal distance error are plotted in Fig. 3.12. As shown in Fig. 3.12a, over 98.8% of the affected pixel clusters can be accurately identified (absolute p_c error less than 1 pixel). In addition, majority of nominal distance error falls in range [0, 1] pixel, and the mean error is around 0.423 pixels (Fig. 3.12b). To reveal the calibration performance for

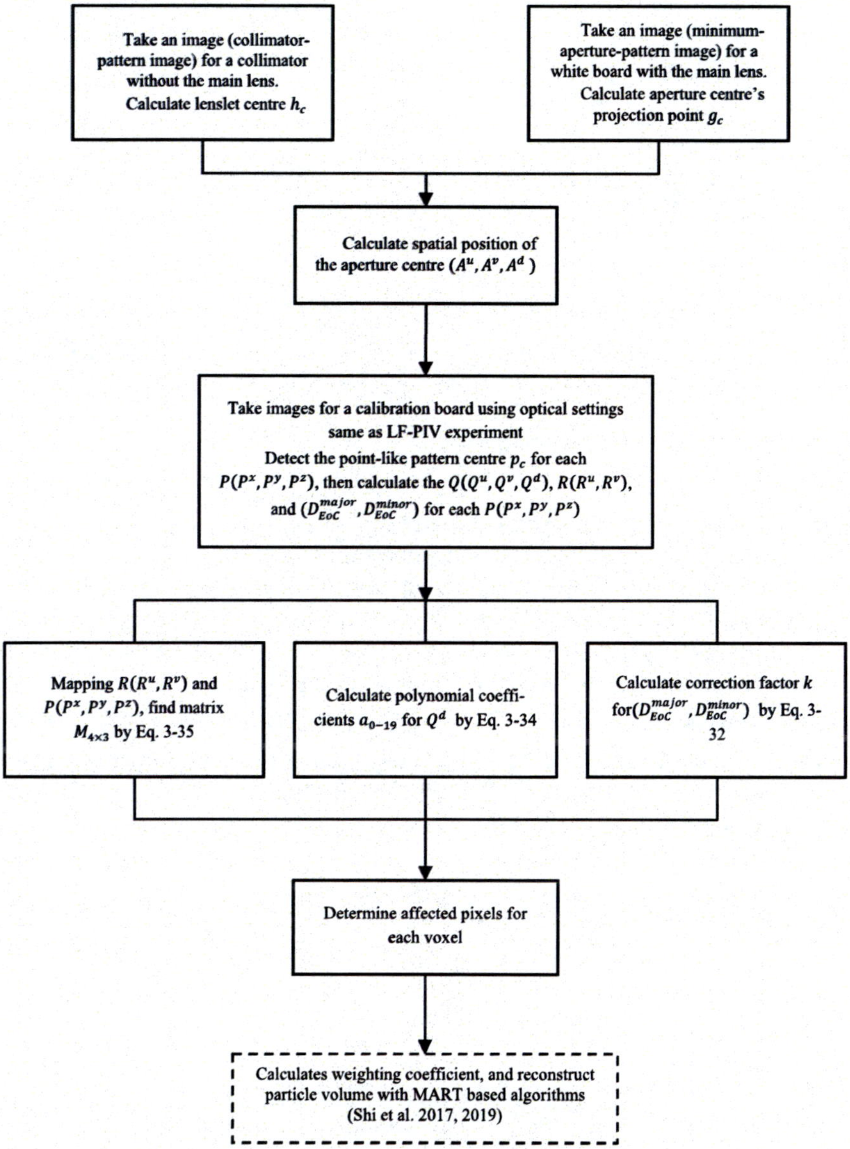

Fig. 3.9 Flowchart of the calibration method for Scheimpflug light-field, the MART reconstruction step is also shown in dash box for completeness. *Source* Shi et al. 2019

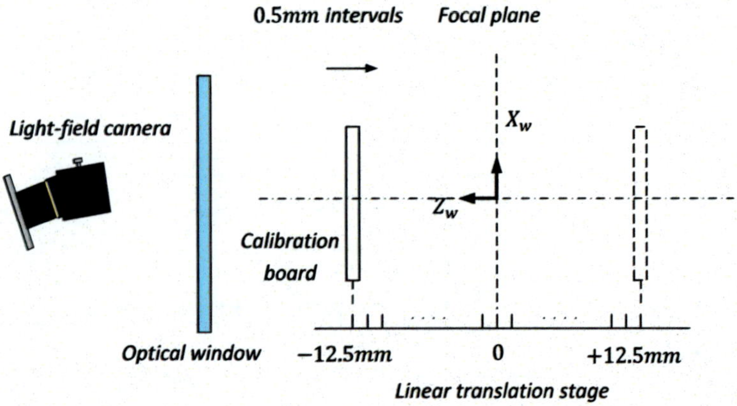

Fig. 3.10 Schematic of the calibration setup

entire measurement volume, the absolute errors in $u-$ and $v-$ direction as well as nominal distance error are plotted along depth direction (Fig. 3.12c). A similar trend as Fig. 3.6b can be observed, calibration error is lower at far sides and reaches a peak near the focal plane. The reasons are also similar, for calibration dots near the focal plane, light rays tend to focus on fewer lenslet, meaning fewer micro-lens are covered. As a result, there are fewer point-like features that could be used to calculate plenoptic disc feature $(R(R^u, R^v)$, $Q\left(Q^u, Q^v, Q^d\right)$ and $\left(D_{\text{EoC}}^{\text{major}}, D_{\text{EoC}}^{\text{minor}}\right))$. The highest error happens at the origin of focal plane $P(0, 0, 0)$, where light rays are focused on only one lenslet and only one equation can be established. In this case, the calibration equations cannot be solved.

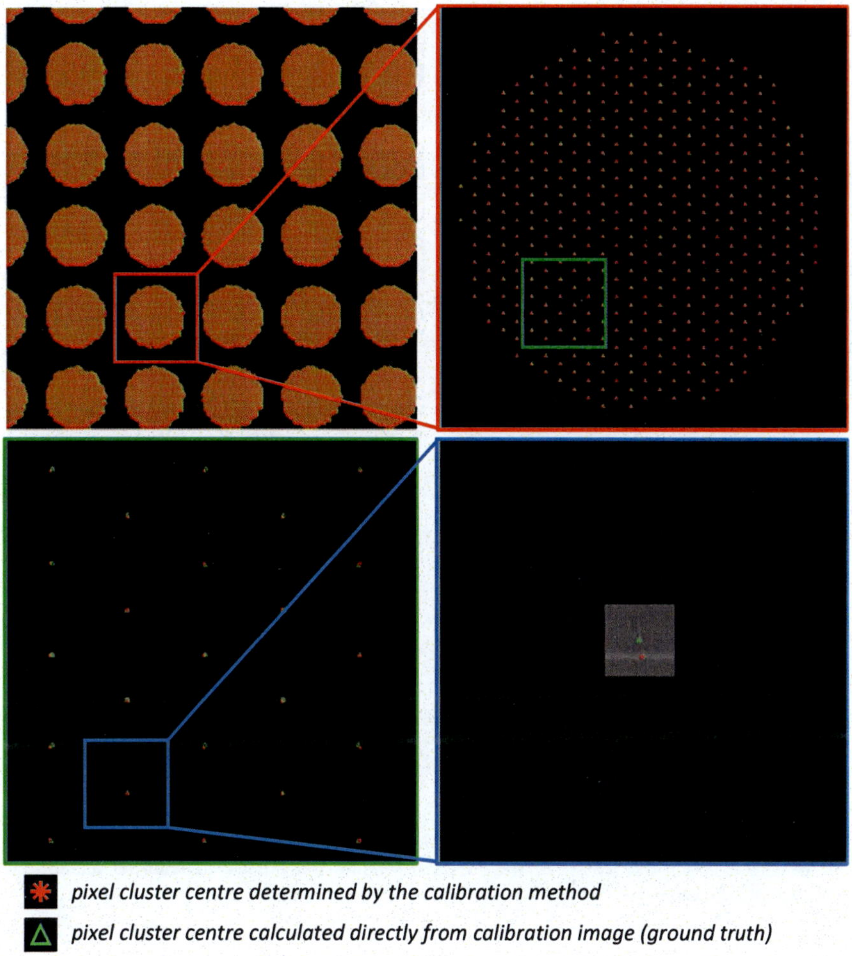

Fig. 3.11 An example of light-field image for one calibration dot, showing together the real centre of affected pixels, together with results from the calibration method

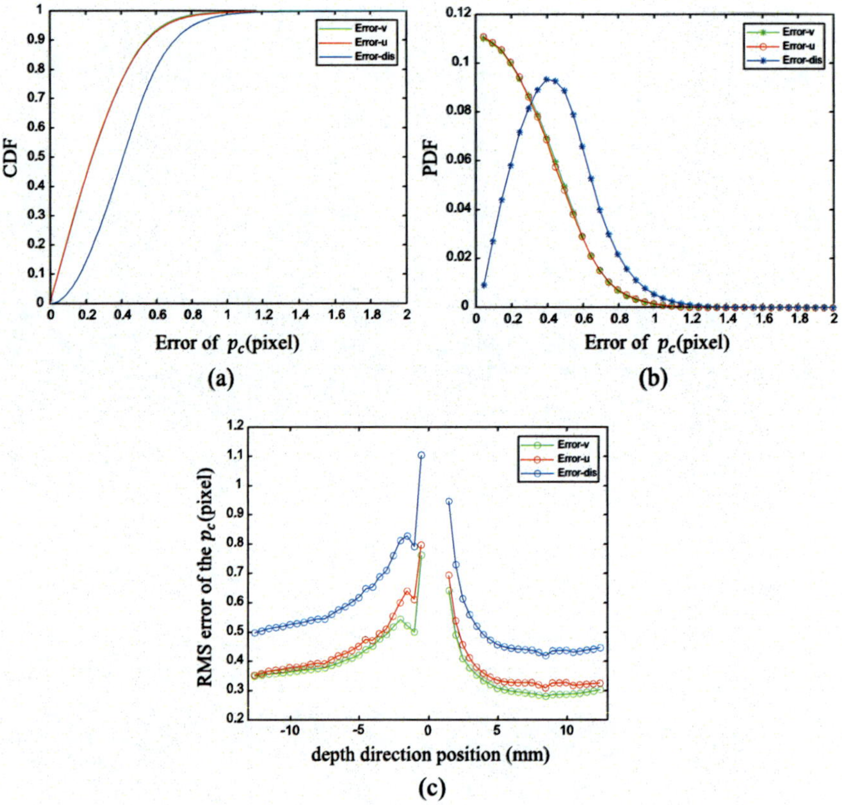

Fig. 3.12 **a** CDF error distribution of the detected point-like feature centre, **b** PDF error distribution of the detected point-like feature centre, **c** error distribution of the detected point-like feature centre along depth direction

References

Evens L (2008a) Depth of Field for the Tilted Lens Plane. Dep Math Northwest Univ 1–20

Evens L (2008b) View camera geometry. Dep Math Northwest Univ 1–109

Merklinger HM (1996) Focusing the view camera. Seabd Print Ltd 5

Ng R (2006) Digital light field photography. Stanford Univ

O'brien S, Trumpf J, Ila V, Mahony R (2018) Calibrating light-field cameras using plenoptic disc features. In: 2018 International conference on 3D vision (3DV). pp 286–294

Shi S, Ding J, New TH et al (2019) Volumetric calibration enhancements for single-camera light-field PIV. Exp Fluids. https://doi.org/10.1007/s00348-018-2670-5

Shi S, Ding J, New TH, Soria J (2017) Light-field camera-based 3D volumetric particle image velocimetry with dense ray tracing reconstruction technique. Exp Fluids. https://doi.org/10.1007/s00348-017-2365-3

Wieneke B (2008) Volume self-calibration for 3D particle image velocimetry. Exp Fluids 45:549–556

Chapter 4
Light-field Particle Image Velocimetry

Shengxian Shi, T. H. New, and J. Soria

Abstract This chapter describes the fundamental principles of Light-field particle image velocimetry (LF-PIV) where accurate detection and reconstruction of seeding particle locations for 3D flow field measurements are of primary concern. Reconstruction of raw light-field particle images and their post-processing based on the dense ray-tracing MART (DRT-MART) approach will firstly be covered, before LF-PIV approach is compared to current Tomo-PIV approach to better understand their unique advantages and disadvantages. A dual light-field camera approach to further improve upon a single light-field camera-based LF-PIV will also be described and discussed here.

Keywords Light-field PIV · MART · Light-field reconstruction · Weighting coefficient · PIV

Introduction

The development of LF-PIV can be better appreciated from a brief history of the authors' earlier works involving other PIV approaches. In earlier investigations, most of the authors' experiments surrounded the use of relatively straight-forward and cost-effective in-house 2D time-resolved particle image velocimetry (TR-PIV) systems. These systems typically comprised of a 532 nm, continuous-wave laser which would be formed into thin laser sheets using appropriate beam steering and sheet-forming

S. Shi (✉)
School of Mechanical Engineering, Shanghai Jiao Tong University, Shanghai 200240, China
e-mail: kirinshi@sjtu.edu.cn

T. H. New
School of Mechanical and Aerospace Engineering, Nanyang Technological University, Singapore 639798, Singapore
e-mail: DTHNEW@ntu.edu.sg

J. Soria
Laboratory for Turbulence Research in Aerospace and Combustion, Department of Mechanical and Aerospace Engineering, Monash University, Melbourne, VIC 3800, Australia
e-mail: julio.soria@monash.edu

S. Shi and T. H. New (eds.), *Development and Application of Light-Field Cameras in Fluid Measurements*, https://doi.org/10.1007/978-3-031-01779-7_4

optics for illumination purposes. A high-speed camera would then be used to capture the seeded flow fields at a sufficiently high frame-rate (and hence, short-time interval Δt) to resolve the transient motions associated with the flow scenarios. After careful calibrations similar to conventional 2D-PIV approaches, the captured particle images could then be post-processed sequentially with the known time interval to arrive at the velocity components based on the typical cross-correlation processing. The main difference between such a TR-PIV approach and conventional 2D-PIV approaches lies in the elimination of the need for a more costly double-pulsed laser and triggering systems. Having said that, the maximum velocity limit for such in-house TR-PIV systems is constrained by the high-speed camera maximum frame-rate, since the minimum time interval is entirely dependent upon it. Furthermore, as the high-speed camera frame-rate increases (with shorter exposure time), the power level required from the continuous-power laser increases as well. Hence, such in-house TR-PIV systems were used for low-to-moderate flow velocities involving water-based experiments, at least for the authors.

The effectiveness of the above in-house TR-PIV systems can be seen in the range of the flow scenarios studied by the authors over the years, especially when information on the transient flow behaviour and quantities are desired. The availability of temporally-resolved velocity field data from the use of these systems also offered a major benefit, in terms of enabling further data-reduction to obtain phase-averaged, Proper Orthogonal Decomposition (POD) and Dynamic Mode Decomposition (DMD) results, take for instance. Provided that sufficiently large number of data is captured over an adequate number of flow cycles (if the flow scenario is cyclical), mean velocity fields and other derived flow quantities are also possible. However, it has to be mentioned that the TR-PIV approach remains 2D in nature and measurements taken along multiple planes continue to be required to provide a more 3D appreciation, especially when attempting to explain the flow physics underpinning the various flow mechanisms. The next logical step will be to utilise Tomo-PIV, though its multi-camera approach is generally more costly and complex, as well as potentially taking up more experimental space. As such, it may not always be the most ideal technique to capture 3D flow measurements non-intrusively.

On the other hand, the idea of using a plenoptic or light-field camera for measurement purposes was explored through a series of systematic studies since the early 2010s, especially by B. Thurow at Auburn University. These developments initially focused on particle-image velocimetry before branching out into other areas such as Background-Oriented Schlieren (BOS), depth measurements, high-temperature measurements, time-resolved measurements and other measurement applications. Whilst the idea of light-field cameras is not new with the first practical implementation demonstrated almost 20 years earlier, where it was shown that light-field cameras could be implemented in a straight-forward manner by putting a layer of MLA slightly ahead of the camera sensor (Adelson and Wang 1992), it was in 2004 and thereafter that these cameras were explored for more commercial usage. The ability to make use of the depth information from light-field cameras to refocus a photo after it has been taken was initially heralded as a breakthrough in photography, though the initial enthusiasm did not catch on with the professional photographers and sales of consumer level light-field cameras did not meet the targets. Unfortunately, they were eventually discontinued but this proved to be step forward towards cost-effective light-field cameras that are well-understood. It was around this point that B. Thurow started to explore the use of light-field cameras for PIV and other flow related measurements (Lynch and Thurowy 2011; Fahringer and Thurow 2012; Lynch et al. 2012; Fahringer and Thurow 2013, 2014, 2015; Thurow and Fahringer 2013; Thomason et al. 2014; Fahringer et al. 2015; Deem et al. 2016; Roberts and Thurow 2017; Klemkowsky et al. 2017; Hall et al. 2018, 2019, amongst others). These studies motivated many further studies, especially those conducted by the authors in terms of furthering the use of light-field cameras in flow measurements and other applications. This chapter will summarise the authors' journey in laying out the design, construction, implementation and post-processing details that pertain towards the use of light-field cameras for measuring a wide range of flow scenarios. In particular, the authors' specific approaches towards an efficient and accurate post-processing algorithm with modern high-speed GPUs will be described here, with a view towards real-time or almost real-time post-processing as the end-goal in mind.

Dense Ray Tracing-based MART 3D Reconstruction of Light-field Particle Images

Light-field particle reconstruction is similar to Tomo-PIV in the sense that they both rely on 2D projections of a tracer particle to reconstruct its 3D image. It is therefore reasonable to expect that MART approach would be a desirable alternative for solving such inverse problem in LF-PIV, as it has been shown to be a very robust one in Tomo-PIV (Scarano 2013). However, the fundamental difference is that rays from a tracer particle in LF-PIV would be recorded by multiple pixels beneath different lenslet (Fig. 2.4), whereas in Tomo-PIV rays from a tracer particle are recorded by multiple cameras from different perspectives. Such a discrepancy would imply a significant challenge when applying MART in light-field particle image reconstruction. For Tomo-PIV, the weighting coefficient could be directly calculated according to the intersection of voxel and pixel line-of-sight, which is normally determined directly from multiple camera calibration (Wieneke 2008, 2018). In contrast, the correspondence of voxel and pixel in light-field imaging is a one-to-multiple mapping (i.e. one voxel affects tens of pixels) as detailed in Chap. 3. As such, the storage of weighting coefficient and direct MART reconstruction are very time-consuming and computationally intensive. For example, the weighting matrix for a $300 \times 200 \times 200$ voxel volume requires 350 GB of storage, even if only non-zero voxel values were to be stored. The reconstruction of such a small volume using the standard MART method takes approximately 1.5h on a 12-core workstation (Fahringer and Thurow 2015).

As with any existing volumetric-based 3D PIV approach, the key towards achieving accurate results depends on accurate reconstruction of the particle images in the 3D space right from the beginning. Seeding particle density for volumetric

PIV tends to be sparse, as it facilitates better particle reconstruction processes. In particular, Atkinson and Soria (2009) demonstrated that the reconstruction process can be significantly accelerated by predetermining the non-zero voxels through the use of a multiplicative line-of-sight (MLOS) approach. This could lead to up to 5.5 times faster particle reconstructions during post-processing as compared to non-MLOS-based approaches, which represent a drastic reduction in the time taken. For volumetric PIV measurements involving a large number of images from multiple cameras, this is a significant breakthrough towards getting 3D PIV measurements faster than ever. However, it should also be noted that the MLOS approach was proposed based on Tomo-PIV technique and cannot be directly used for LF-PIV without modifications. For Tomo-PIV, camera calibration information can be used to work out the line-of-sight of a pixel, where the non-zero voxels may then be identified subsequently by multiplying the corresponding pixels. The situation for LF-PIV is far more complicated however, as illustrated in Fig. 2.4. The figure depicts the lines-of-sight for a point light source (or illuminated tracer particle) located at the focal plane, at some distances dz away from the focal plane and dy away from the camera axis, for a light-field camera. As can be seen from the depiction, the line-of-sight of a pixel is highly sensitive towards the particle location and non-zero voxels must be identified through inverse ray-tracing to find the concerned pixels. This can however be done if the central light ray for each discretised section of the main lens were to be raytraced and provide information for particle reconstruction. To demonstrate this principle, Fig. 4.1 illustrates the proposed approach using an example of five pixels for each lenslet of the MLA, where the red region represents a voxel in the measurement region. Through this approach, pixels associated with each voxel can be ascertained and a straight-forward multiplication of their values can be used to differentiate the non-zero voxels. Such a technique is actually similar to that utilised by MLOS approach. Once the non-zero voxels have been identified, the intensity of the voxel can subsequently be calculated using:

$$E\left(X_j, Y_j, Z_j\right)^{k+1} = E\left(X_j, Y_j, Z_j\right)^k \left(\frac{I(x_i, y_i)}{\sum_{j \in N_i} w_{i,j} E\left(X_j, Y_j, Z_j\right)^k}\right)^{\mu w_{i,j}} \tag{4.1}$$

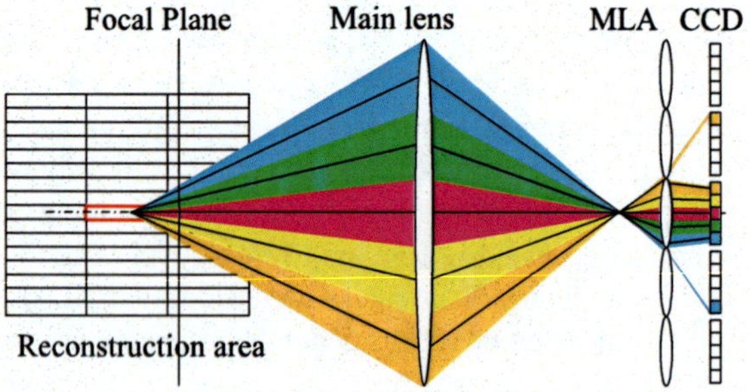

Fig. 4.1 Illustration of proposed approach where a hypothetical number of five pixels are associated with each lenslet

where $E(X_j, Y_j, Z_j)$ is the intensity of the *j*th voxel; $I(x_i, y_i)$ is the intensity of the *i*th pixel, which is known from the captured light-field image; and w_{ij} is the weighting coefficient, which is the contribution of light intensity from the *j*th voxel to the *i*th pixel value.

As with particle reconstruction used for Tomo-PIV, a weighting coefficient will need to be determined, though it will be very different for LF-PIV due to the different line-of-sight principles between Tomo-PIV and LF-PIV. Shi et al. (2016) proposed that the weighting coefficient can be calculated from ray-tracing technique, where it will be used to relate between the voxel, lenslets and pixels. In fact, the weighting coefficient, w, is calculated from two different parts, w_1 and w_2, and a result of their product (i.e. $w = w_1 \times w_2$). In particular, w_1 is calculated from the overlapping area between the light beam and lenslet as shown, whereas w_2 is calculated from the overlapping area between the light beam and the pixels. To better explain the rationale and principles behind the weighting coefficient w, assume that there are 5 × 5 pixels behind each lenslet as shown in Fig. 4.2a. Subsequently, discretised light bundles are traced from the voxel under consideration to the microlens plane. Take for example

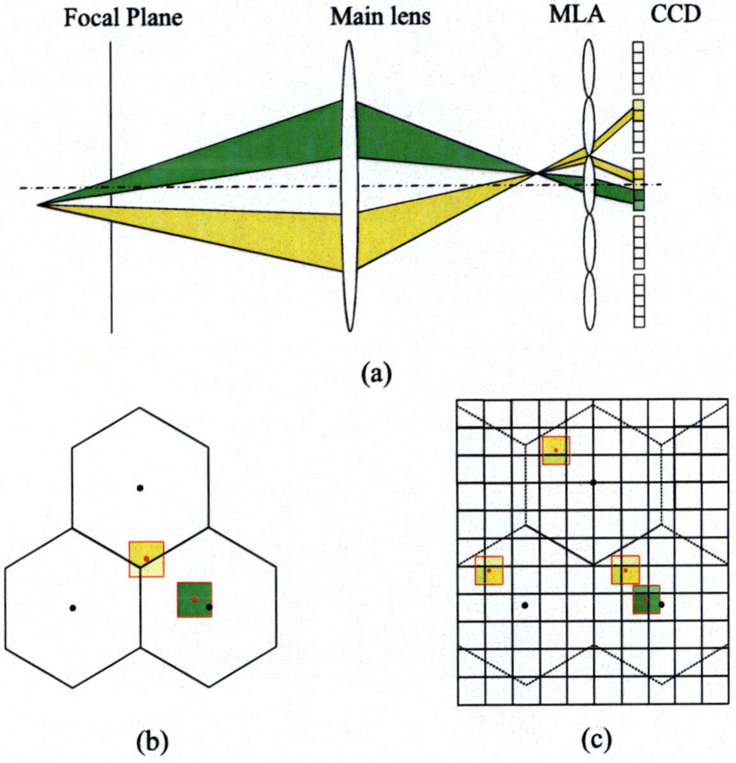

Fig. 4.2 Schematics depicting weighted ray tracing principles on **a** how ray tracing can be used to locate the affected lenslet and pixels beneath, **b** overlapping area between light ray and lenslet and **c** overlapping area between light ray and pixel

the lower yellow light bundle in Fig. 4.2a, ray tracing will enable the determination of its location, which in turn allows its overlap area with three adjacent lenslets to be calculated for w_1. As physical pixel geometries are almost always squarish, the projection of the light bundle on the microlens plane is illustrated using squares. With w_1 taken care of, procedures to obtain w_2 will now be elaborated. To do that, consider the continuing tracing of the light bundle from the microlens to the image sensor plane, such that the position of the centre light ray is determined as shown in Fig. 4.2c. Once that is ascertained, w_2 may then be calculated as the overlap area between the affected pixels and the projected sub-light bundle. To demonstrate the efficacy of this weighting coefficient for LF-PIV, Fig. 4.3a shows a synthetically generated light-field image whilst Fig. 4.3b and c shows the weighting coefficient distributions (depicted as an image) calculated by the preceding procedures. It can be observed that the present procedures are able to obtain significantly more accurate weighting coefficient distributions, and hence form the basis of what the authors termed as "dense ray tracing MART" or DRT-MART reconstruction technique. The

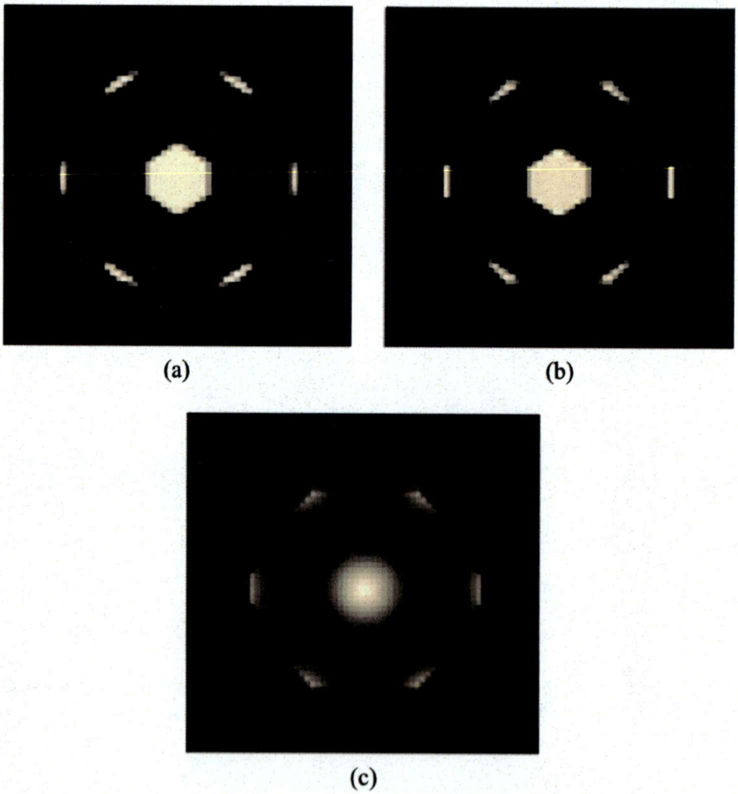

(a) (b) (c)

Fig. 4.3 **a** Synthetic light-field image of a point light, **b** weighting coefficient calculated by ray tracing method and **c** weighting coefficient calculated by sphere–cylinder intersection algorithm (Shi et al. 2016) next few sections will compare the performance and efficacy of DRT-MART against MART in terms of some common considerations when reconstructing 3D particle images.

Reconstructed 3D Particle Elongations

Just like in Tomo-PIV, LF-PIV imaging tends to produce elongation of particle images along the optical axis. This is in fact a known issue with Tomo-PIV even though multiple cameras are used (Soria and Atkinson 2008; Scarano 2013). For typical LF-PIV, this is caused by the use of a single-camera-based approach. To better understand this elongation phenomenon and compare between the MART and DRT-MART approaches, a study was conducted based on synthetic volumetric 3D particle images, where the particles were randomly distributed. These synthetic images then were post-processed by both MART and DRT-MART approaches to reconstruct the light-field particle images systematically. Subsequently, the diameters of reconstructed particles were determined at locations where voxel intensity was less than two standard deviations away from the maximum voxel values, which defines a diameter of 3 voxels for an ideal particle image. Note that both MART and DRT-MART approaches used here should incur the same computational time, such that the accuracy levels achieved within the same time can be compared directly. As such, the iteration numbers for MART and DRT-MART are 23 and 400 respectively. Increasing the iteration number for MART approach further is impractical as that will drastically increase the computational time. Perhaps more importantly, results to be presented later also show that the DRT-MART approach is superior over MART approach even if the latter is allowed to undergo 400 iterations. For a closer look, Fig. 4.4 shows the reconstructed 3D particle images obtained by DRT-MART and MART approaches with 400 iterations, as well as MART approach with 23 iterations. It can be observed from the figure that DRT-MART approach leads to smaller elongation of the reconstructed particle within a shorter reconstruction time. This significantly better performance is due to the dense ray-tracing eliminating the zero voxels before the reconstruction stage. In contrast, MART approach does not do this and instead, reconstructs both the non-zero voxels and the neighbouring zero voxels, as its weighting coefficient is unable to ascertain the exact affected lenslet and pixels.

To quantify the above, probability distribution functions (PDF) of the reconstructed particle diameter in the three primary directions (i.e. x, y and z) were determined for both DRT-MART and MART approaches and compared in Fig. 4.5. Whilst it is clear that particle elongations in the x and y directions are not that significant for both approaches, DRT-MART approach nevertheless leads to generally smaller particle diameters of approximately 2–4 pixels in these directions, whilst MART approach produces slightly large particle diameters of about 5–7 pixels. The discrepancy becomes much larger however in the z direction (i.e. depth direction along the optical axis for the present analysis) are compared. In this case, DRT-MART

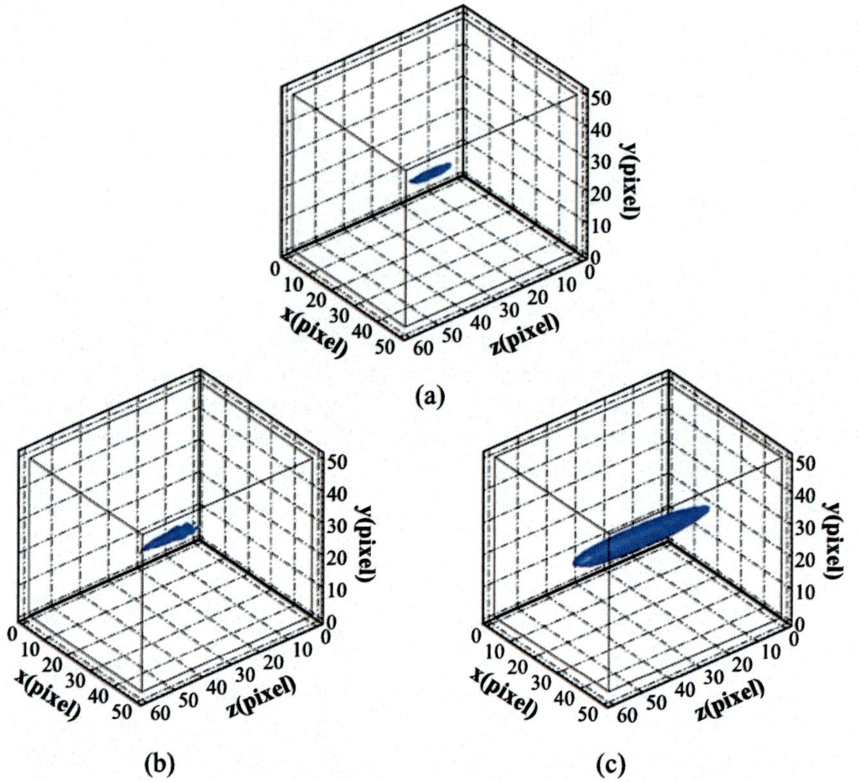

Fig. 4.4 Reconstructed 3D particle images obtained by **a** DRT-MART and **b** MART approaches with 400 iterations, as well as **c** MART approach with 23 iterations

achieves significantly smaller particle diameters of about 10–25 pixels, whereas MART approach produces about 35–45 pixel particle diameters. The outcome is clearly much more severe for the MART approach in the depth direction.

Reconstruction Quality and Speed

Reconstruction quality is known to be influenced by iteration number, relaxation factor and particle density, amongst others, and this section will take a systematic look at how each of these three factors influences the subsequent accuracy levels associated with DRT-MART and MART approaches. Firstly, a typical 3D volume, albeit a small one, of 0.1 particle per microlens (PPM) was used to generate synthetic light-field particles images. These images were subjected to reconstructions by DRT-MART and MART approaches, where a range of iteration numbers and relaxation factors were used to study how the latter will impact the reconstruction quality of the

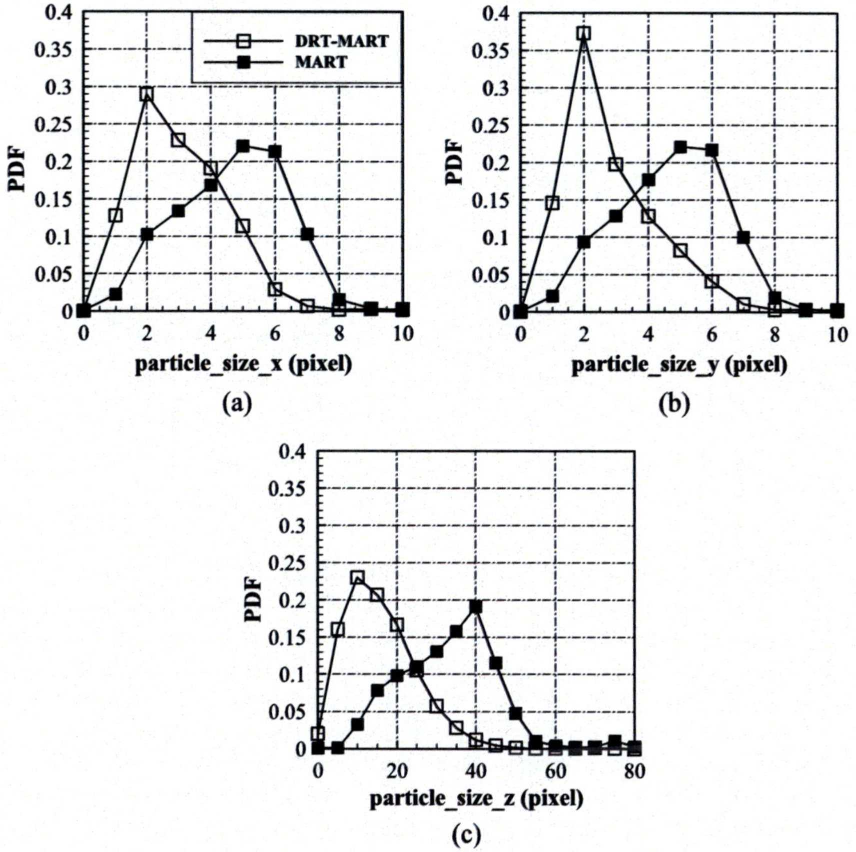

Fig. 4.5 Probability density functions of reconstructed particle diameters in the **a** x, **b** y and **c** z directions. Note that z direction is also the depth direction along the camera axis

two approaches. The reconstructions were carried out by discretising the volume with pixel-voxel ratios of 1:1 in the x and y directions, as well as 10:1 in the z direction, respectively. A Q_{Recon} factor was used to quantify the reconstruction quality (Elsinga et al. 2006) and defined as

$$Q_{\text{Recon}} = \frac{\sum E_1(x, y, z) E_0(x, y, z)}{\sqrt{\sum E_1^2(x, y, z) \times E_0^2(x, y, z)}} \tag{4.2}$$

where $E_0(x, y, z)$ is the exact voxel intensity approximated by a Gaussian distribution with three voxel diameter, and $E_1(x, y, z)$ is the voxel intensity of the reconstructed particle image.

Figure 4.6 shows the variations in the reconstruction quality level with respect to changes in the iteration number and relaxation factor μ, after the preceding analysis.

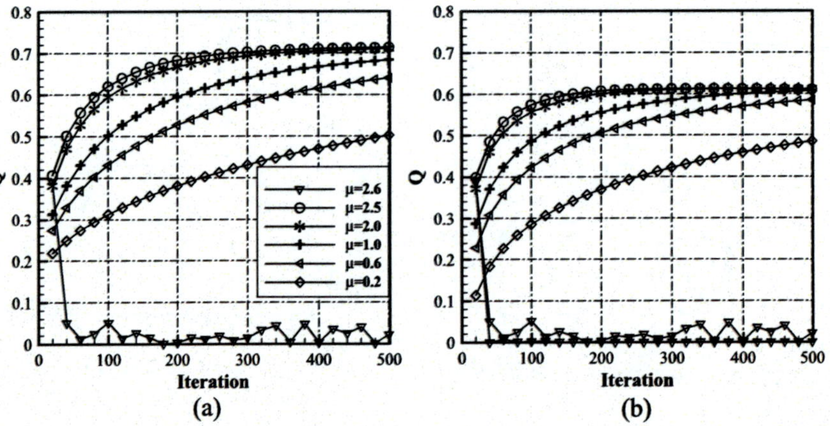

Fig. 4.6 Impact of iteration number and relaxation factor upon the reconstruction quality for **a** DRT-MART and **b** MART approaches

Firstly, it can be appreciated from these results that DRT-MART approach is able to reach consistently higher reconstruction quality levels than MART approach, the reason being DRT-MART approach produces less elongated particles with more Gaussian-like voxel intensities. Secondly, the relaxation factor and iteration number for both DRT-MART and MART approaches in LF-PIV need to be larger than that for Tomo-PIV for comparable reconstruction quality levels. This is due to the fact that a more significant number of pixels are affected by every voxel in LF-PIV, as opposed to just several pixels per voxel in Tomo-PIV. In fact, even more pixels will be affected in LF-PIV if the voxel is located further away from the focal plane. This can be appreciated in Fig. 4.7 where the plots show how the maximum voxel intensity varies as the depth-of-field changes, and how the use of DRT-MART and MART approaches affect it. Regardless of the exact approach used, it can be seen that the maximum voxel intensity levels do not vary very much when close to the focal plane (i.e. 80–110 voxel) and much of the iterations beyond the 20th iteration were actually going towards reconstruction efforts further away from the focal plane. This leads to the understanding that the calculations will have to consider the number of pixels and their relative contributions towards the final voxel intensity, and explain why LF-PIV generally needs more iterations and larger relaxation factors to reach high reconstruction quality levels.

As mentioned earlier, volumetric PIV measurement techniques such as Tomo-PIV and LF-PIV tend to make use of relatively low particle densities within the measurement volumes for satisfactory 3D particle reconstruction outcomes. Theoretically speaking, high particle densities should be used as they could provide smaller interrogation volumes and hence higher measurement resolutions, and the use of low particle densities seems to be counter-intuitive. However, it had been discovered in earlier Tomo-PIV studies that a higher particle density leads to more ghost particles that prevent high reconstruction quality (Elsinga et al. 2006; Scarano 2013), even

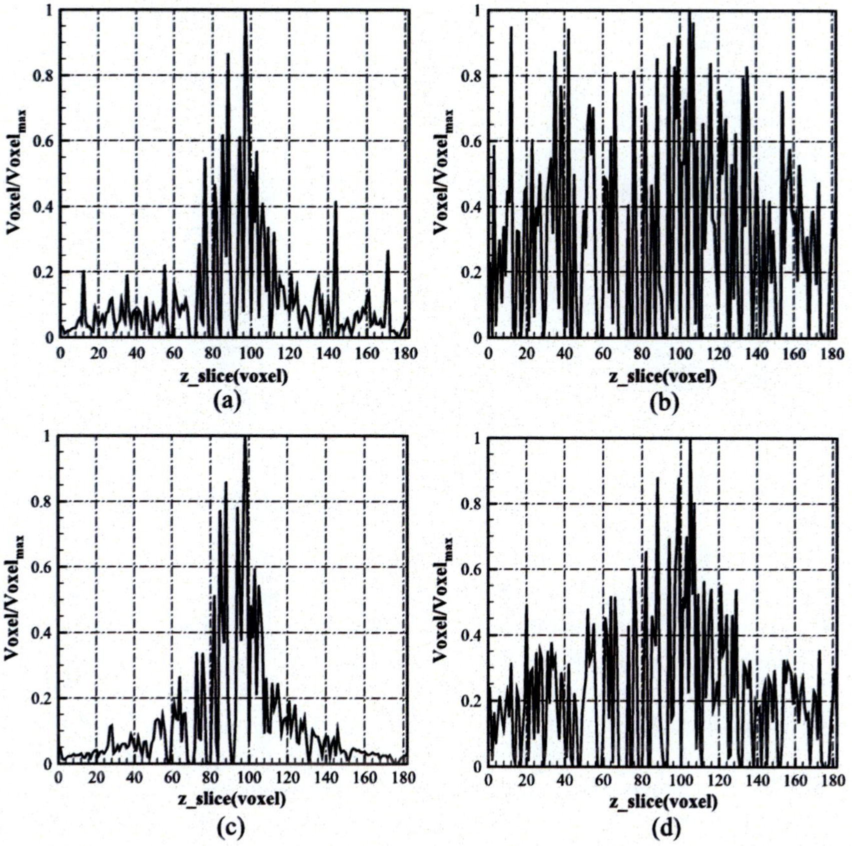

Fig. 4.7 Variations of maximum voxel intensity with depth-of-field changes (i.e. in the *z* direction) for DRT-MART approach with **a** 20 iterations and **b** 400 iterations, as well as MART approach with **c** 20 iterations and **d** 200 iterations

though multiple cameras were utilised. LF-PIV faces the same problem, especially since it typically makes use of a single light-field camera. To have a better appreciation of the issue, synthetic light-field particle images were once again generated but with different particle densities, where they were reconstructed using both DRT-MART and MART approaches with 400 and 200 iterations respectively. Other than that, a 2.5 relaxation factor and pixel-voxel ratios of 1:1 (in x and y directions) and 10:1 (in the z direction) were maintained throughout. The outcomes of this particular analysis are shown in Fig. 4.8, and it is quite clear that DRT-MART approach attains better reconstruction quality level than MART approach at the same particle density. Furthermore, the reconstruction quality deteriorates as the particle density increases, regardless of either approach, similar to Tomo-PIV. Note that no ghost particles are observed during the particle reconstruction stage for DRT-MART and MART approaches, due to information made available by the multiple perspectives offered

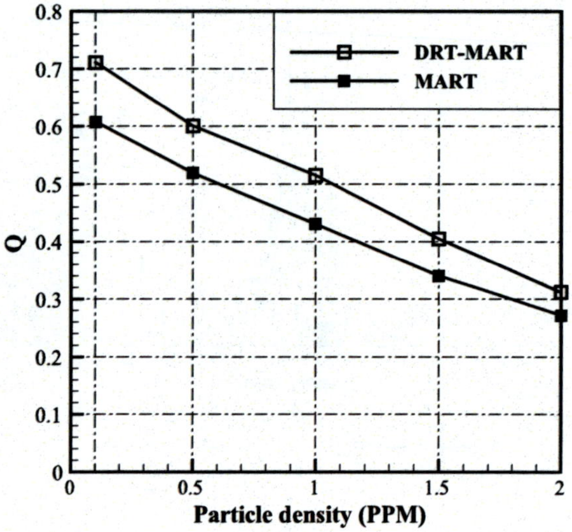

Fig. 4.8 A comparison of the impact upon reconstruction quality level due to particle density for DRT-MART and MART approaches

by the microlens (Ng 2006). To confirm this, particle reconstruction was extended in the z direction in both positive and negative directions to include volumes which did not have any seeding particles and this was done on 30 synthetic light-field particle images. The reconstruction results were summed up and the voxel intensity levels were then averaged along the x–z plane. These results were plotted and presented in Fig. 4.9 for a closer look at how they vary along the pertinent direction/plane. As Fig. 4.9a shows, the summed-up voxel intensity clearly shows zero intensity beyond the region where particles exist (i.e. blue region) and that the voxel intensity

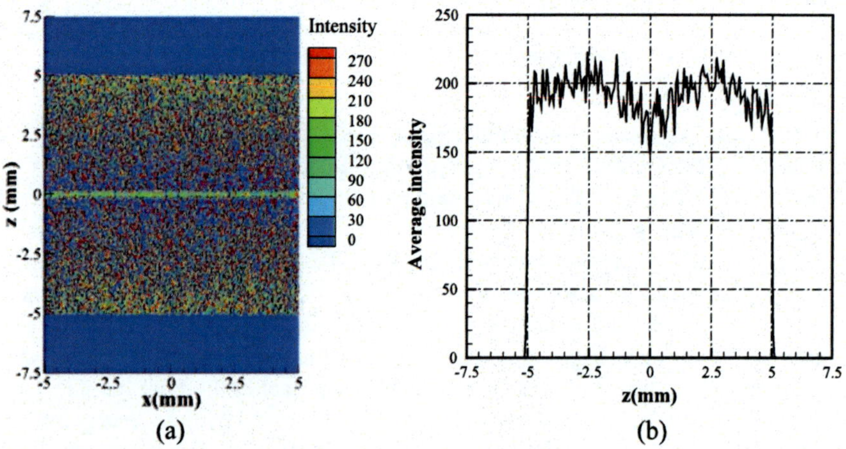

Fig. 4.9 Variations in the **a** sum of 30 reconstructed light-field particle images taken along the x–z plane, and **b** average voxel intensity level along the z direction

is generally lower along $z = 0$ (i.e. greenish line). The latter can be attributed to the reconstructed particles around the focal plane occupying more voxels (elongation effect) with reduced intensity, which is due to lower resolution close to the focal plane as highlighted earlier on. This can be better appreciated in greater detail in Fig. 4.9b, where the average voxel intensity taken in the z direction along $x = 0$. The dip in the voxel intensity close to the focal plane (i.e. $z = 0$) and the abrupt drop to zero levels beyond the seeded region can be easily discerned.

Last but not least, another comparison of interest between DRT-MART and MART approaches here will be their computational speeds. To do that, the same synthetic light-field particle images were used for reconstruction by DRT-MART and MART approaches, where 400 and 200 iterations were used for the former and latter respectively. Furthermore, the particle density was varied as part of the computational efficiency comparison, since it is expected that increasingly higher particle density will lead to greater computational loads. The results are presented in Fig. 4.10, where the computational time for MART approach was non-dimensionalised by that for DRT-MART approach and plotted as a function of particle density. As the figure shows, DRT-MART approach is faster than MART approach until about 2PPM. However, note that 1PPM is typically the upper particle density limit for such 3D particle reconstructions, it remains clear that DRT-MART continues to enjoy a significant speed advantage over MART approach by being about 4 times faster at that particle density.

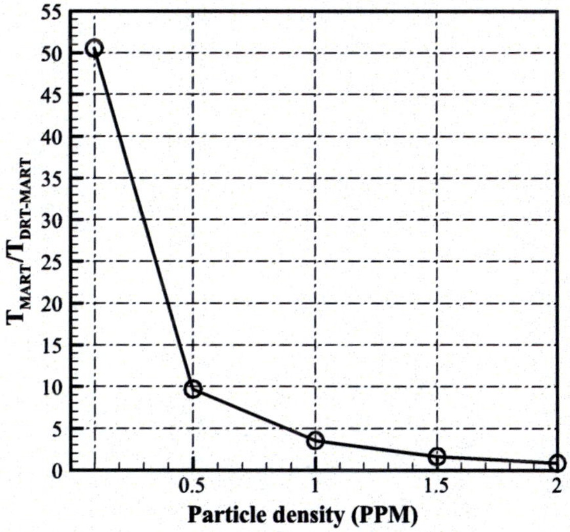

Fig. 4.10 Computational time taken by MART approach relative to DRT-MART approach, as a function of particle density

Reconstruction Accuracy Under Unsteady Flow Conditions

The previous section discussed upon the reconstruction efficacy of DRT-MART and MART approaches with respect to the particle seeding condition and reconstruction details, using light-field particle images of randomly dispersed particles. Whilst it demonstrates the potential of LF-PIV, it is not representative of the largely unsteady flow fields that are typically encountered in fluid dynamics and aerodynamics research. Therefore, this section will examine the two reconstruction approaches based on more complex flow fields, such that a better grasp of their accuracy levels under these circumstances can be attained. For this particular analysis, a variety of unsteady oscillatory flow fields were used to ascertain the relative performance of DRT-MART and MART approaches. To be more specific, the flow fields can be divided into two types:

Type A where the oscillatory motion is only in the *x*-direction:

$$\begin{cases} u(x,y,z) = 25e^{i(k_x x + k_y y + k_z z)} \\ v(x,y,z) = 0 \\ w(x,y,z) = 0 \end{cases} \tag{4.3}$$

And Type B where the oscillatory motion is only in the *z*-direction:

$$\begin{cases} u(x,y,z) = 0 \\ v(x,y,z) = 0 \\ w(x,y,z) = 25e^{i(k_x x + k_y y + k_z z)} \end{cases} \tag{4.4}$$

Each of these two types was further divided into four different configurations and Table 4.1 shows the details of these configurations. Synthetic light-field particle images were subsequently generated according to these flow scenarios for further analysis. Of particular interest will be the effects of pixel–voxel ratio, iteration

Table 4.1 Details of the various oscillatory flow scenarios

Case	k_x	k_y	k_z	L
A1	$2\pi/\omega L_x$	0	0	1820 pixel (L_x)
A2	$2\pi/\sqrt{2}\omega L_{xy}$	$2\pi/\sqrt{2}\omega L_{xy}$	0	$\sqrt{2} \times 1820$ pixel (L_{xy})
A3	$2\pi/\sqrt{2}\omega L_{xz}$	0	$2\pi/\sqrt{2}\omega L_{xz}$	$\sqrt{2} \times 1820$ pixel (L_{xz})
A4	0	0	$2\pi/\omega L_z$	1820 pixel (L_z)
B1	$2\pi/\omega L_x$	0	0	1820 pixel (L_x)
B2	$2\pi/\sqrt{2}\omega L_{xy}$	$2\pi/\sqrt{2}\omega L_{xy}$	0	$\sqrt{2} \times 1820$ pixel (L_{xy})
B3	$2\pi/\sqrt{2}\omega L_{xz}$	0	$2\pi/\sqrt{2}\omega L_{xz}$	$\sqrt{2} \times 1820$ pixel (L_{xz})
B4	0	0	$2\pi/\omega L_z$	1820 pixel (L_z)

Note that $\omega = 0.25, 0.5, 1.0$ and 2.0 were used as part of the different combinations number and velocity gradient on the hypothetical measurement accuracy on DRT-MART and MART approaches. Seeding density of 0.5PPM was used, whilst pixel–voxel ratios (PVR) of 1, 2 and 3 were adopted for x and y directions. As for reconstructions, DRT-MART and MART approaches made use of 400 and 200 iterations respectively, with a relaxation factor of 2.5. 3D multi-grid cross-correlations (Soria 1996; Atkinson and Soria 2009) were used to process the reconstructed particle images with an overlapping ratio of 0.75, as well as initial and final interrogation volumes of 320 × 320 × 32 voxel and 160 × 160 × 16 voxel, respectively.

To compare how well DRT-MART and MART approaches perform here, displacement errors between the known flow field and measured results for Case A1 in the x and y directions are determined and their PDF presented in Fig. 4.11a. It can be seen that DRT-MART approach performs better than MART approach at PVR = 1 and 2, whilst there is little difference between PVR = 1 and 2 for DRT-MART approach. Although PVR = 1 could in theory offer smaller interrogation volume and hence better measurement resolution, the number of particles within a single interrogation volume may not meet the requirement for accurate cross-correlation. Equally important, using PVR = 2 rather than 1 will actually accelerate the DRT-MART reconstruction by up to 4 times. Moving on to the z-direction, the displacement error PDF for Case B1 is now presented in Fig. 4.11b, where PVR = 5, 10 and 20 in z direction were used whilst PVR = 1 was maintained in the x and y directions. Results show that DRT-MART achieves best performance at PVR = 10, the latter of which produces a 2-voxel diameter particle that conforms well with an idealised 3D Gaussian-type geometry. In contrast, the significantly poorer performance put up by MART approach is due to the resulting much larger reconstructed particle sizes.

Next, the effects of iteration number on the displacement errors are considered and Fig. 4.12 shows the relationships between the RMS values of the displacement errors and iteration number in both the x and z directions. In the x direction, about

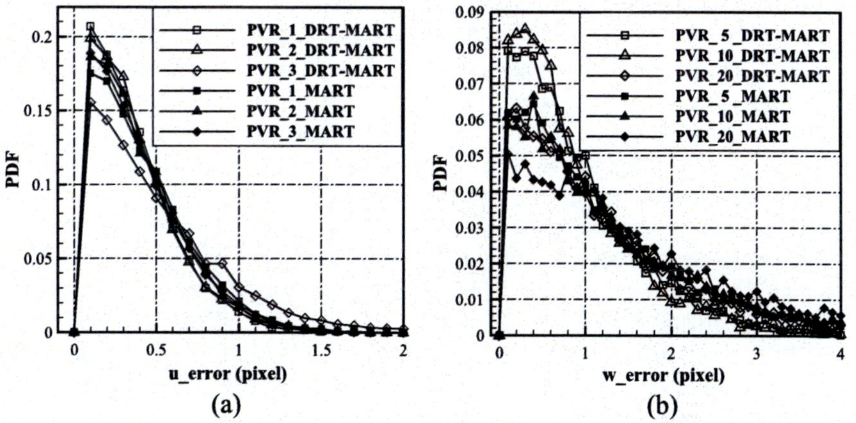

Fig. 4.11 Displacement error PDF in the **a** x and y directions, as well as **b** z direction under different PVR values

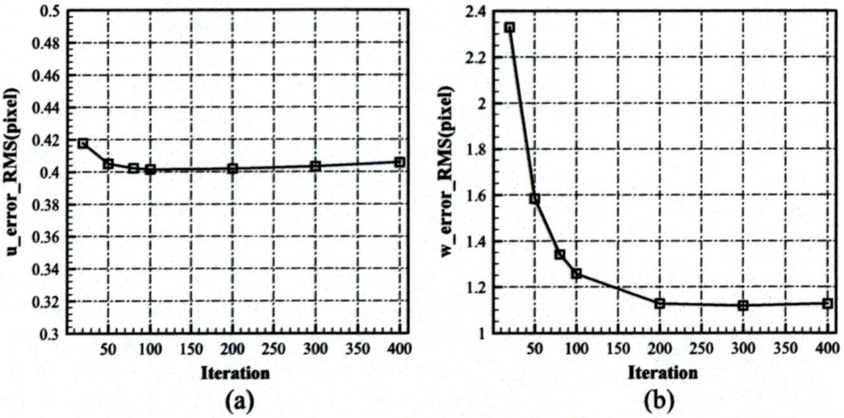

Fig. 4.12 Relationships between RMS velocity errors and iteration number in the **a** x and **b** z directions for DRT-MART approach

50 iterations lead to satisfactorily low measurement errors in the x (and y) direction, whilst approximately 200 iterations will be needed to reduce the measurement errors in the z direction to a sufficiently low level. This is consistent with the earlier finding that a higher iteration number is necessary to achieve satisfactory reconstruction of particles that are located further away from the focal plane.

Last but not least, attention is now turned towards understanding how the presence of significant velocity gradients will impact upon the measurement accuracy level. This will be of significant interest as most engineering flows involve turbulent shear flow scenarios and therefore, a more extensive comparison based on all 8 configurations will be shown here. All test cases were based on synthetic light-field particle images generated at 0.5PPM, relaxation factor of 2.5, PVR = 1 in the x and y directions, and PVR = 5 in the z directions. For a more consistent comparison, similar computational times were allocated to DRT-MART and MART approaches, which lead to 200 and 23 iterations for the former and latter respectively. Subsequently, 3D multi-grid cross-correlations with 75% overlapping ratio, as well as initial and final interrogation volumes of 320 × 320 × 32 voxels and 160 × 160 × 16 voxels, were used respectively. Similar to the preceding comparisons, displacement error PDF results are presented in Figs. 4.13 and 4.14 for all test cases to compare and evaluate the two different reconstruction approaches.

As expected and shown in Fig. 4.13, higher measurement accuracies will be achieved for low and moderate velocity gradient scenarios (i.e. $\omega = 0.25$ and 0.5) for both reconstruction approaches. However, it remains clear that DRT-MART approach is still discernibly better than MART approach. It is also noteworthy to point out that it is only when the u velocity component varies along the z direction in Case A4 that the performance of both reconstruction approaches suffers a significant drop in measurement accuracy. This can be appreciated from the fact that the measurement resolution is lower along the depth and hence z direction. As for Fig. 4.14, similar

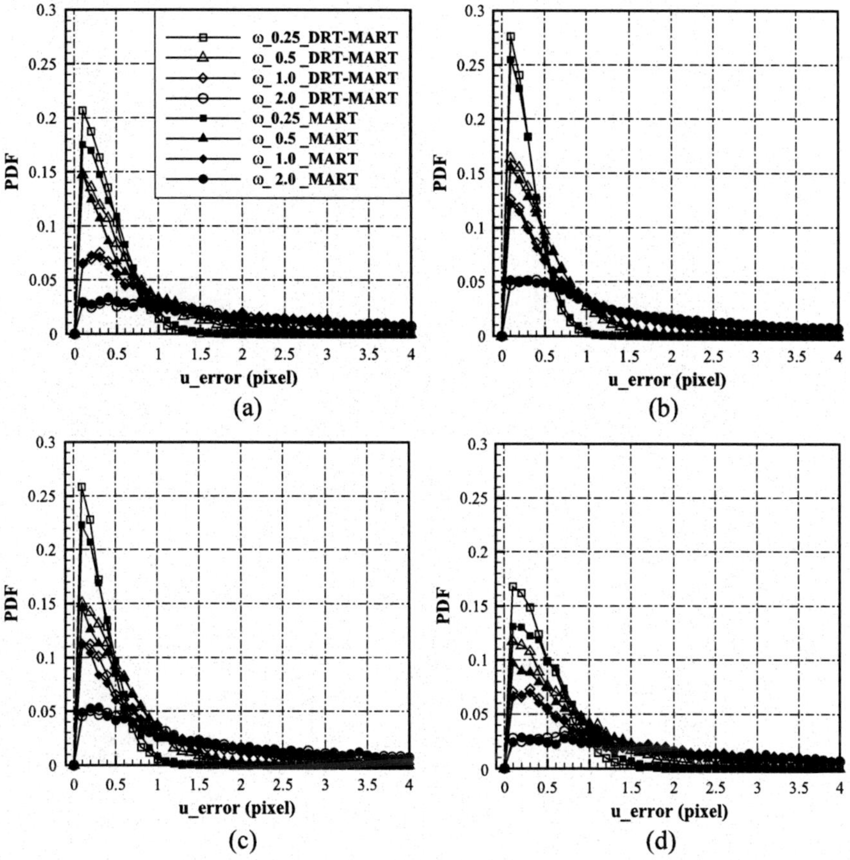

Fig. 4.13 Comparison of displacement error PDF in the x direction between DRT-MART and MART approaches for **a** case A1, **b** case A2, **c** case A3 and **d** case A4

outcomes can be observed as well, where the lowest measurement accuracy is for Case B4 where the w velocity component varies in the z direction (and depth direction). In this case, measurement accuracy is far worse than the situation for Case A4. These observations reinforce the notion that the light-field camera axis should not be aligned in the direction where the most dominant velocity component exists, but instead perpendicular to it for higher measurement accuracy levels.

Experimental Validations

Up to this point, all testing had been done based on synthetic light-field particle images and as much as theoretical oscillating velocity fields make things more realistic, they are still not based on real-world flow scenarios. Nevertheless, all the testing

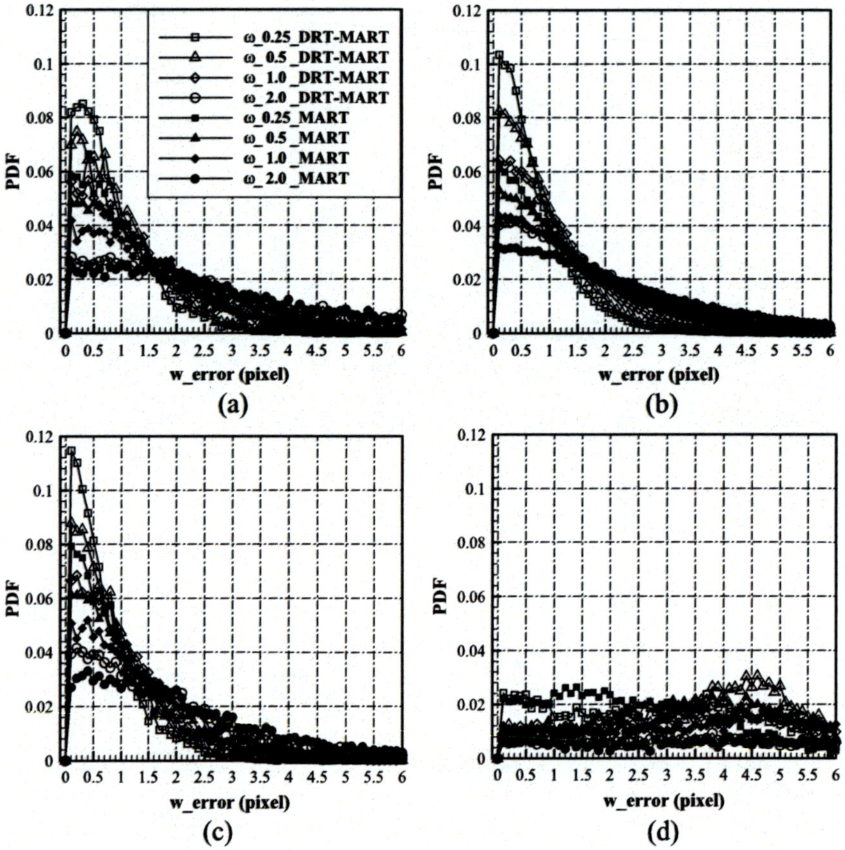

Fig. 4.14 Comparison of displacement error PDF in the z direction between DRT-MART and MART approaches for **a** case B1, **b** case B2, **c** case B3 and **d** case B4

thus far had provided confidence and much needed understanding on the advantages and disadvantages of the proposed DRT-MART approach. So, undertaking LF-PIV measurements and making use of DRT-MART on captured light-field particle images to arrive at the 3D flow fields of real-world flow scenarios will be the key "litmus test". To do that, one of the first validation experiments was conducted on a canonical laminar, incompressible round water jet flow at a Reynolds number of 2000. The design of the experimental setup was generally similar to those adopted by earlier studies on various jet flow phenomena (New and Tsai 2007; New and Tsovolos 2012; Shi and New 2013; Long and New 2015, 2016, 2019) and its operations will hence only be briefly described here. With reference to Fig. 4.15, water from a small reservoir was channelled into the jet apparatus by a centrifugal water pump, where it passed through a diffuser, honeycomb, three layers of fine screens and a contraction chamber. Water would subsequently exhaust from a $D = 20$ mm diameter round nozzle into a large Plexiglas water tank filled with quiescent water. To ensure a

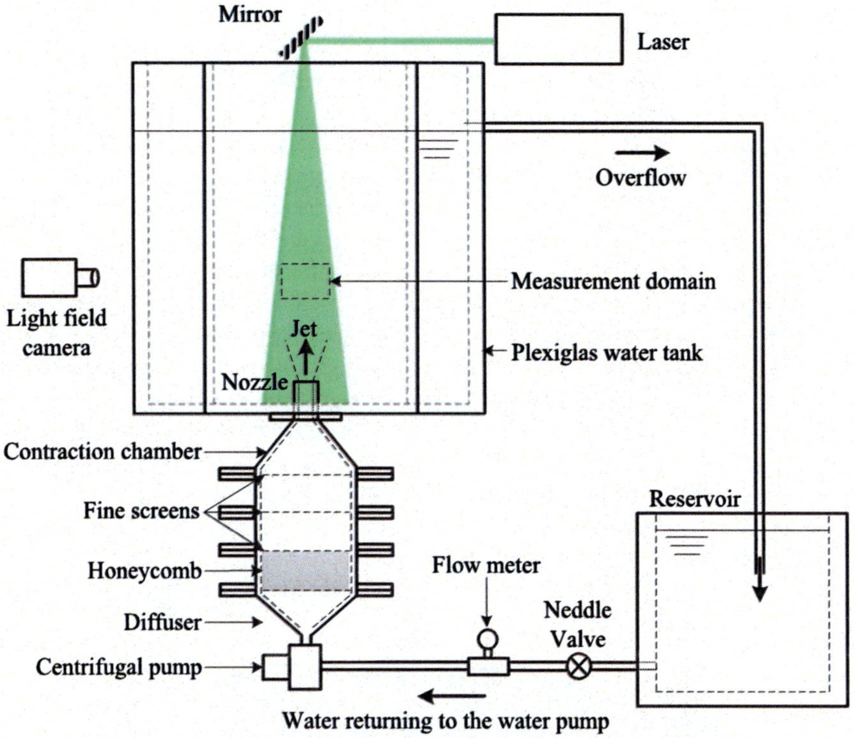

Fig. 4.15 Schematics of the experimental setup and flow circuit used for the experimental testing

constant static pressure head, excess water would be channelled out of the water tank via an overflow pipe and back into the small reservoir, thus closing the flow circuit. Flow velocity adjustments were done using a needle valve and monitored using an electromagnetic flowmeter. To ensure high fidelity during the measurements, the measurement volume was restricted to $1.9D \times 1.3D \times 0.5D$ located at $2.25D$ above the nozzle exit. The aim was to capture the regular and coherent vortex roll-ups along the jet shear layer due to Kelvin–Helmholtz hydrodynamic instabilities. 20 μm polyamide seeding particles of 1.03 g/cm^3 density were dispersed and circulated throughout the flow circuit and water tank at about 0.4PPM. 10 mm thick laser sheets were produced by a 200 mJ/pulse, 532 nm wavelength Nd:YAG laser to provide volumetric illuminations and an in-house light-field camera (Shi et al. 2016) was used to record the light-field particle images. Additionally, main lens aperture was 4 and magnification factor was −0.95. At the same time, a 2 ms time interval was used to ensure that the one-quarter particle displacement rule for satisfactory cross-correlations was adhered to.

Before reconstructing the 200 captured light-field particle images, the global background was subtracted from them to better filter out the zero-voxels. Subsequently, DRT-MART and MART approaches were used to reconstruct the 3D particle images based on 400 and 23 iterations respectively. The reconstruction involved 3300 × 2200 × 182 voxels, used a relaxation factor of 2.5, and PVR = 2 in the x and y directions, as well as PVR = 10 in the z direction. 3D multi-grid cross-correlations were used to process the reconstructed particle images, where 75% overlapping ratio, initial and final interrogation volumes of 320 × 320 × 64 voxels and 160 × 160 × 32 voxels, as well as a 3-point × 3-point median filter to reject spurious vectors. Before touching upon the results however, one issue associated with practical experimentation using LF-PIV will need to be highlighted. Figure 4.16 shows the distributions of the voxel intensity levels along the x–z plane and z direction, similar to what had been before for synthetic light-field particle images and shown in Fig. 4.9 earlier. Unlike what had been observed for Fig. 4.9 earlier, the voxel intensity levels along and close to the focal plane are actually much higher than at locations further away. Further investigation revealed that the Gaussian distribution of the laser beam (and hence laser sheet) intensity is responsible for this phenomenon, where the situation is exacerbated by the significantly lower intensity levels even further away from the focal plane. Note that calibration errors for the microlens and optical aberration would have contributed towards the lower intensity levels as well.

Figure 4.17 shows the post-processed results produced by both DRT-MART and MART approaches and it should be recalled that the aim was to capture the coherent

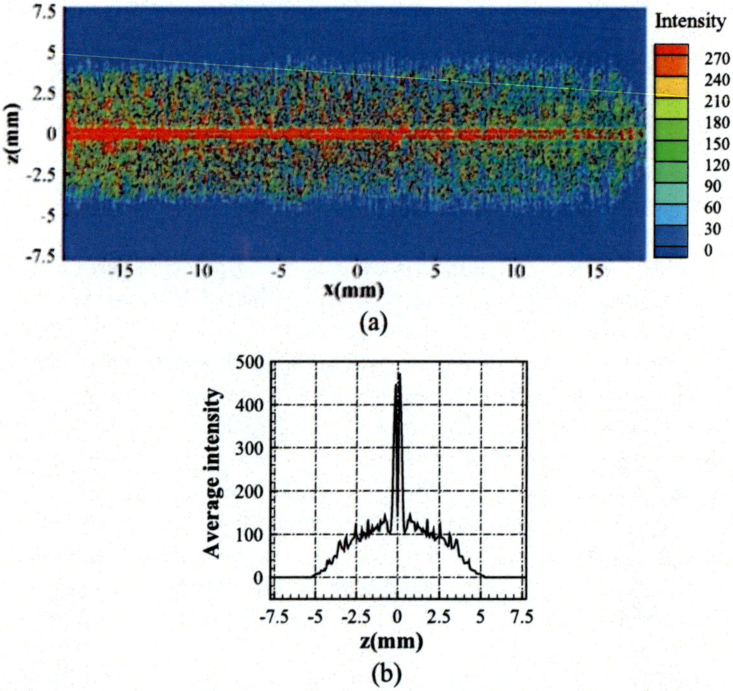

Fig. 4.16 Variations in the **a** sum of 30 reconstructed light-field particle images taken along the x–z plane, and **b** average voxel intensity level along the z direction

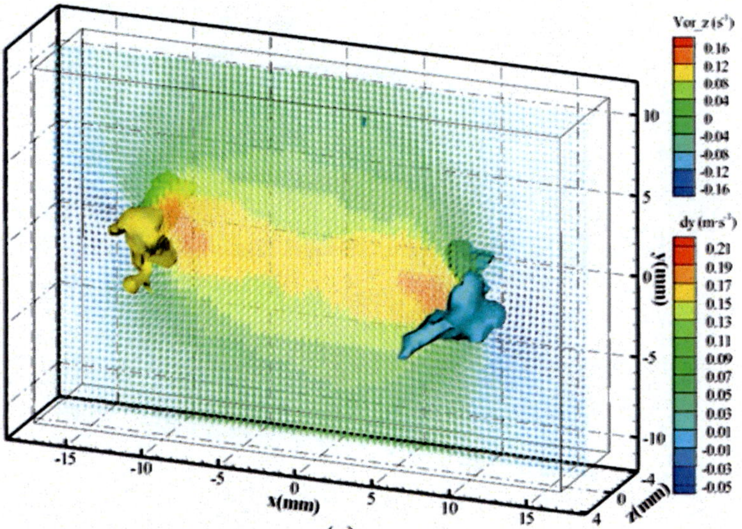

(a)

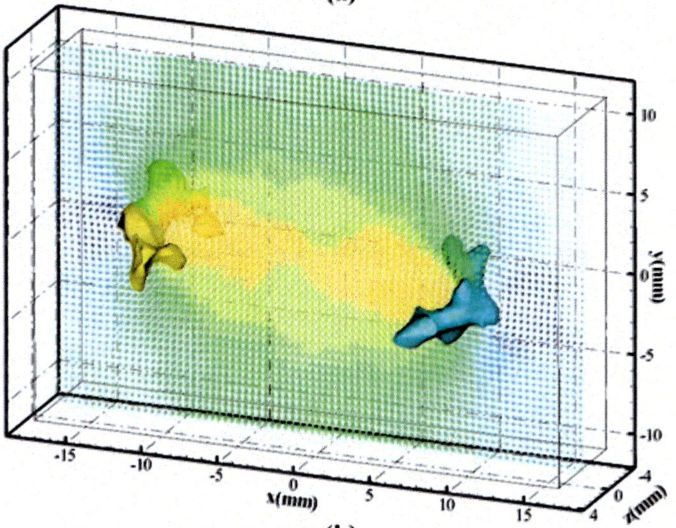

(b)

Fig. 4.17 Instantaneous velocity vector field and vorticity isosurfaces generated by **a** DRT-MART and **b** MART approaches

vortex-roll-ups along jet shear layer. From that perspective, it would appear that the present LF-PIV configuration, as well as DRT-MART and MART approaches were able to resolve the 3D flow behaviour within the measurement volume well. By most accounts, the vortex roll-up structure and behaviour captured by the two different approaches appear to be very similar from that 3D orientation, barring some minor differences. However, it should be noted that the velocity vector field is along the

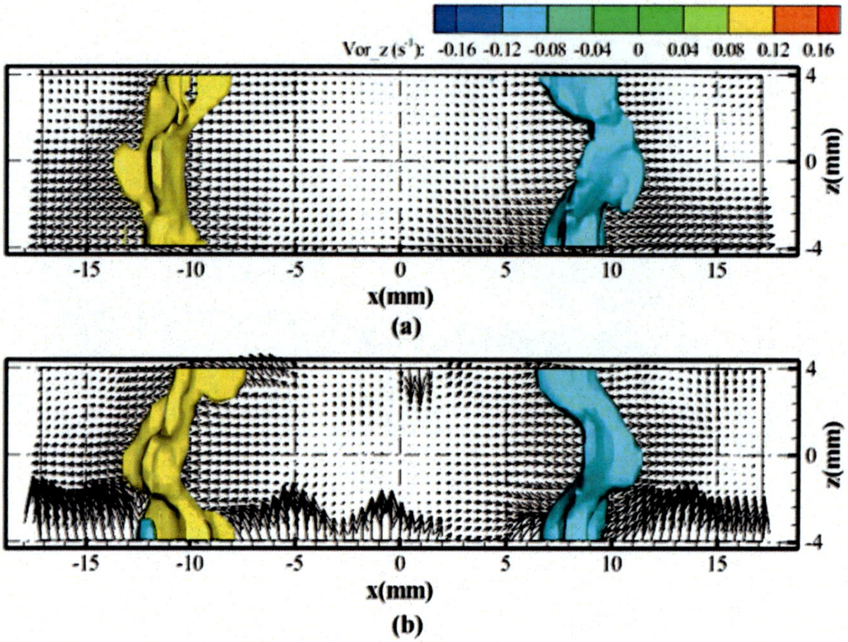

Fig. 4.18 Top-views of the instantaneous 3D flow fields corresponding to Fig. 4.17

x–y plane and $y = 0$ and little information on the z direction can be inferred from there. To inspect closer, Fig. 4.18 shows the top-views of the results corresponding to Fig. 4.17. From this orientation, it becomes clear that MART approach produces significant number of erroneous w velocity vectors at the furthest distances away from the focal plane located at $y = 0$. In contrast, DRT-MART approach produces reasonable w velocity vector distributions in the same regions. This discrepancy seen in the experimental results produced by DRT-MART and MART approaches attested to the findings arising from the earlier study based on synthetic light-field particle images. In fact, this further reinforces the notion that utilising synthetic particle images continues to be very useful in testing out novel PIV techniques and post-processing procedures.

Comparison with Tomo-PIV

Tomo-PIV is currently the most popular volumetric approach when it comes to 3D flow measurements and it will be very useful to compare the preceding single camera LF-PIV approach with conventional multi-camera Tomo-PIV approach. In particular, since one of the biggest benefits of LF-PIV approach has been its potential to make use of a single light-field camera instead of multiple cameras, it will be instructive to compare their accuracy levels. In addition, it will also be interesting to find out at what light-field camera Tomo-camera pixel ratio will a single camera LF-PIV be able to achieve similar accuracy levels as with Tomo-PIV. This is especially important since advances in imaging sensor technology meant that sensor pixel numbers will continue to increase rapidly and may one day be sufficiently dense and cost-effective that the convenience offered by LF-PIV drives a higher adoption rate. To find out more, synthetic light-field and tomographic particle images were used to study the impact of camera number in Tomo-PIV, as well as pixel resolution ratio between light-field and tomographic cameras, upon the overall accuracy levels of the reconstructed 3D flow fields. Once that had been accomplished, further comparisons were conducted based on actual experiments on laminar, incompressible jet flows for a better understanding of the practical experimental challenges and implications associated with the two different volumetric 3D PIV approaches.

Synthetic Particle Image Generation and Analysis

Before the details of how the synthetic particle image were generated and analysed by LF-PIV and Tomo-PIV approaches, it is important to firstly highlight the inherent differences between these two approaches and how they will affect the generation of synthetic particle images. For Tomo-PIV, it is well known that the number of cameras and particle density have strong influences upon its accuracy levels (Elsinga et al. 2006; Atkinson and Soria 2009). This is different from LF-PIV where the situation is more complex, depending on how the MLA is configured. As introduced in Chapter 2, a light-field camera where the MLA is located at one focal length distance away from the imaging sensor will produce the highest angular resolution possible. On the other hand, different spatial and angular resolutions will result if the distance between the MLA and imaging sensor deviates from that (Georgiev and Intwala 2006; Lumsdaine and Georgiev 2009) in unfocused light-field cameras. For the purpose of volumetric velocity measurements, it is preferable to have a higher angular resolution than spatial resolution, since it means the ability to gather more information on the out-of-plane particle displacements. It had been shown in earlier studies that LF-PIV approach is heavily influenced by the pixel-microlens ratio (PMR) and a higher MLA resolution can better handle higher particle densities that lead to higher spatial resolution. In fact, the larger the number of pixels associated with each microlens, the higher the angular resolution. Coupled with the desired higher MLA resolution, it is unsurprising that the capability of LF-PIV approach increases with the imaging sensor pixel resolution.

For a more consistent comparison between LF-PIV and Tomo-PIV, a light-field camera to Tomo-camera pixel ratio defined as

$$\mathrm{LTPR} = \frac{(p_x \times p_y)_{\mathrm{LF\text{-}PIV}}}{(p_x \times p_y \times N_c)_{\mathrm{Tomo\text{-}PIV}}} \tag{4.5}$$

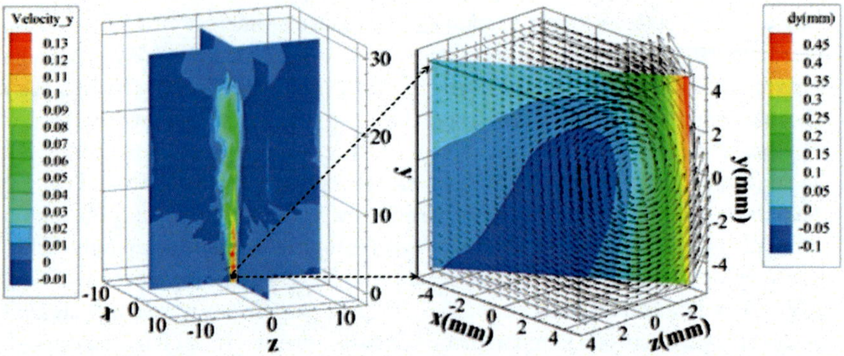

Fig. 4.19 A snapshot of the Re = 2500 incompressible jet flow simulated using DNS. The inset shows the details of a vortex roll-up along the jet shear layer

was used to quantify theoretically how much times higher a light-field camera resolution needs to be over than all the Tomo-camera resolutions added together. On top of that, it also shed some light upon the cost of a LF-PIV setup relative to a Tomo-PIV setup. One thing that needs to be highlighted is that synthetic light-field images generated for LF-PIV and Tomo-PIV approaches here will be based on the most favourable particle density associated with each of them, rather than being the same throughout. The reason for this is that Tomo-PIV approach is able to handle higher particle densities than LF-PIV due to its use of multiple cameras (Scarano 2013; Fahringer et al. 2015; Shi et al. 2016, 2017). However, it will be desirable to compare the two approaches when they are optimised for a given field-of-view, so that their full capabilities can be better compared here.

For the present comparison, particular attention was paid towards how variations in PMR, LTPR and the number of cameras for Tomo-PIV will impact the relative advantages of LF-PIV approach. In particular, results from Direct Numerical Simulations (DNS) of a Re = 2500 incompressible jet flow were used to generate the synthetic particle images for LF-PIV and Tomo-PIV, as shown in Fig. 4.19. Note that the measurement volume was taken at one jet diameter above the nozzle exit, so that vortex roll-ups (see inset of Fig. 4.19) would be captured in the results. With reference to the figure, orientation of the hypothetical light-field camera is arranged such that its optical axis is along the z direction, whilst its imaging sensor is facing the x–y plane. Briefly describing, to begin generating the synthetic light-field particle images, particles are randomly dispersed in the very first frame first, before they were been displaced based on the DNS jet flow result for a fixed time interval that satisfy the one-quarter rule for PIV particle displacements. With the particle locations determined for each synthetic particle images, their corresponding light-field particle images would be generated by using 5 million rays per particle. These rays were then traced from each particle through the main lens and MLA according to principles laid out in Georgiev and Intwala (2006) and Shi et al. (2016).

For LF-PIV, two different hypothetical light-field camera resolutions of 800 × 800 pixels and 1600 × 1600 pixels with PMR = 7, 14 and 28 were tested. The MLA were also assumed to comprise of hexagonal lenslets for higher resolutions. As for Tomo-PIV, each camera was assumed to have a resolution of 160 × 160 pixels with a possibility of 4, 6 and 8 camera combinations. The preceding test cases produce LTPR of 3.13, 4.17, 6.25, 12.5, 16.67 and 25, covering a significant range of ratios. Generally speaking, the aperture between the main lens and microlenses should be matched to optimise the resolution. However, maintaining this optimal condition will lead to changes to other parameters when the operating condition changes. For instance, a change in PMR results in a change in the lenslet size and the focal length of the latter will need to be adjusted to ensure aperture number is maintained. Figure 4.20 depicts several scenarios to demonstrate how different PMR will lead to changes to the separation distance between the MLA and imaging sensor. Readers can refer to Table 1 of Shi et al. (2018) for the different combinations of parameters used here for both LF-PIV and Tomo-PIV for better clarity. Last but not least, 0.5PPM was used as it was found to produce optimal results (Shi et al. 2016). With the preceding settings, synthetic light-field particle images could then be generated and some of them are shown in Fig. 4.21a–c.

Moving to Tomo-PIV synthetic particle image generation, they were generated based on equally spaced cameras (i.e. 4, 6 and 8) along a semi-circle arc in the *x*–*z* plane as shown in Fig. 4.22. The pinhole camera model (Tsai 1986) was used to

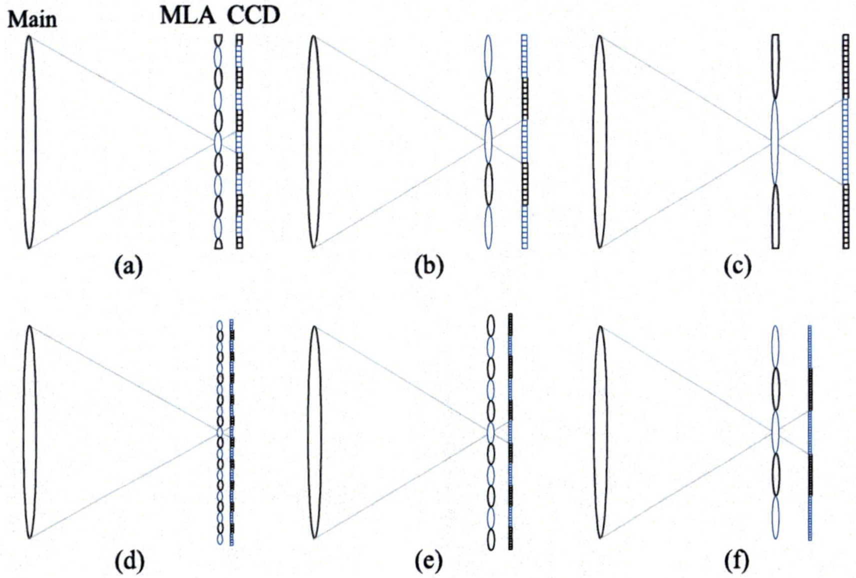

Fig. 4.20 Schematics showing how changes to PMR will lead to changes to the separation distance between the MLA and imaging sensor. **a** PMR = 7, LTPR = 3.13, 4.17, 6.25; **b** PMR = 14, LTPR = 3.13, 4.17, 6.25; **c** PMR = 28, LTPR = 3.13, 4.17, 6.25; **d** PMR = 7, LTPR = 12.5, 16.67, 25; **e** PMR = 14, LTPR = 12.5, 16.67, 25; **f** PMR = 28, LTPR = 12.5, 16.67, 25

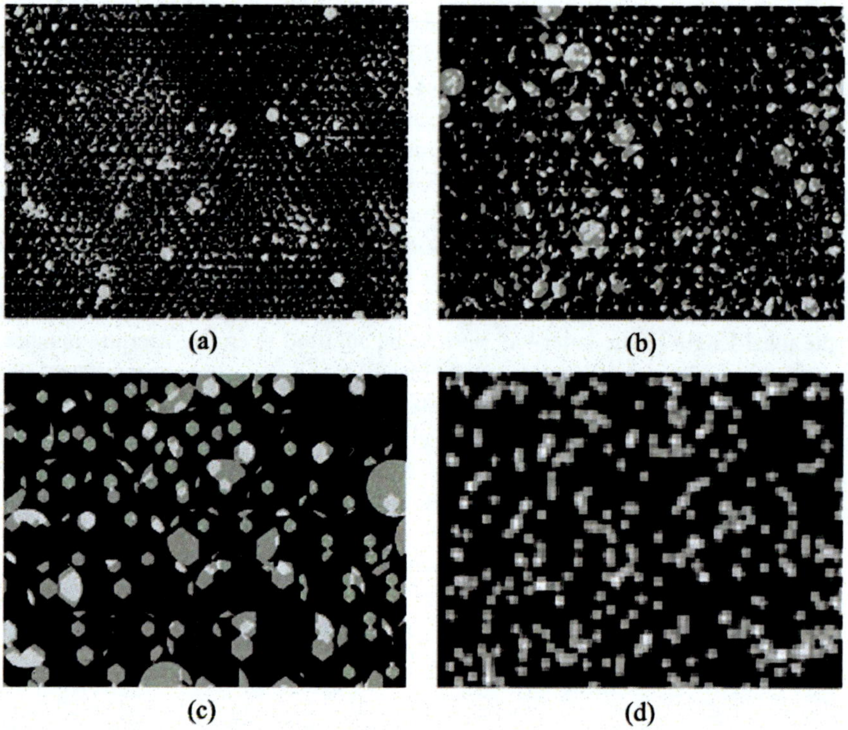

Fig. 4.21 Example synthetic particle images generated for reconstruction by LF-PIV and Tomo-PIV approaches **a** LF-PIV (LTPR = 3.13, 4.17, 6.25, PMR = 7, 0.5PPM); **b** LF-PIV (LTPR = 3.13, 4.17, 6.25, PMR = 14, 0.5PPM); **c** LF-PIV (LTPR = 3.13, 4.17, 6.25, PMR = 28, 0.5PPM); **d** Tomo-PIV (0.05PPP)

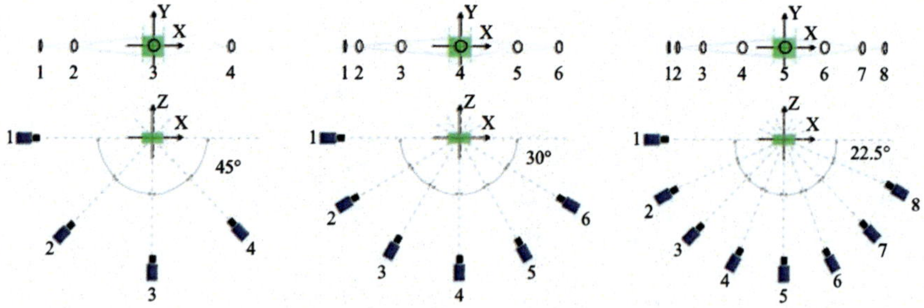

Fig. 4.22 Multi-camera Tomo-PIV configurations used in present comparison

calculate the projected views of a particle in 3D space upon the 2D imaging sensor with a focal length of 85 mm. Image and object distances of 93.4 and 946.5 mm were used as well, for a 0.075 mm/pixel magnification factor. For multiple cameras in Tomo-PIV, projections of all the particles upon imaging sensors were determined using the camera matrix and particle centre locations. To achieve a reasonable particle diameter size of about 3 pixels, a Gaussian distribution was applied for particle images and this was repeated for all cameras used. This would produce the first instance of the synthetic particle images to be studied by Tomo-PIV here. Similar to what was adopted for LF-PIV, DNS results were then used to displace the particles according to the simulated velocity fields over a selected time interval and the earlier procedures were repeated again for subsequent synthetic particle images. To post-process the synthetic particle images using LF-PIV and Tomo-PIV techniques, DRT-MART (Shi et al. 2017) and MLOS-SMART (Atkinson and Soria 2009) approaches were used respectively. In the case for Tomo-PIV, a 1:1 pixel-voxel ratio was used. In contrast, pixel-voxel ratios used in the x, y and z direction were set to 2, 2 and 10, in accordance to earlier findings that show their suitability for LF-PIV approach. To arrive at the 3D velocity fields, 3D multi-grid cross-correlations similar to what had been used earlier on were used and their details can be found in Table 2 of Shi et al. (2018). Note also that the particle centres were calculated based on peak centroid method, instead of Gaussian peak fitting.

The first comparison focused on how much the reconstructed particle centre locations deviated from the simulation results and the PDF of the reconstruction errors in all three directions for both LF-PIV and Tomo-PIV approaches (Fig. 4.23). For Tomo-PIV, it is not surprising to see that increasing the camera number and decreasing particle density led to improvements in the reconstruction accuracy levels. As earlier studies had shown (Elsinga et al. 2006; Scarano 2013), elongations of reconstructed particles and ghost particles can be mitigated by adopting wider viewing angles (i.e. more cameras) and lower particle densities. On the other hand, LF-PIV approach is more affected by camera resolution, as increasing it produces much higher reconstruction accuracy levels in the z and y directions. Furthermore, reconstruction accuracy in the z direction is determined more by PMR (Shi et al. 2016) and to a certain extent, the LTPR. For instance, a larger LTPR leads to higher reconstruction accuracy since it will improve MLA resolution, which in turn leads to the light rays being captured by more lenslets/pixels and better reconstruction of the particle z direction displacement.

Next, discrepancies between the LF-PIV and Tomo-PIV velocity field results were compared with the DNS jet flow results through the RMS errors for each velocity components individually, as shown in Fig. 4.24. Starting with the low-LTPR LF-PIV case, it should be noted that whilst a higher PMR produces better depth resolution, it also leads to larger measurement errors. Furthermore, a higher PMR means that the lenslet physical size has to increase and therefore, a lower number of total lenslets that can be deployed. What this implies is that maximum acceptable particle density will decrease, with significant impact upon the minimum interrogation volume that can be used for cross-correlations and hence measurement accuracy. When the LF-PIV camera resolution is increased significantly as shown in the figure, the measurement

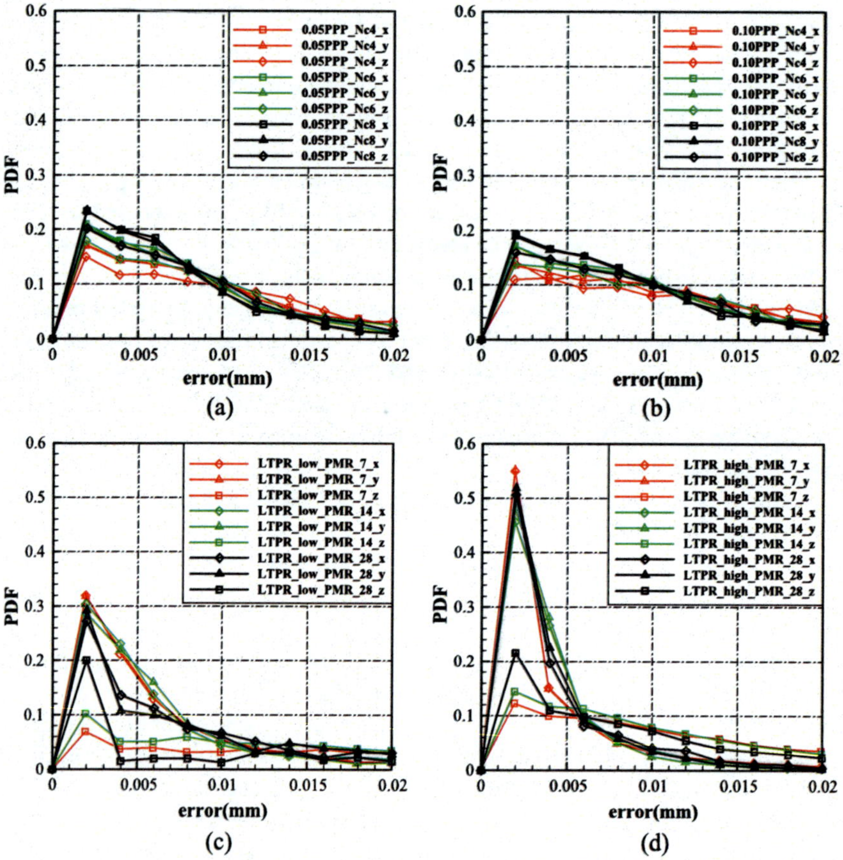

Fig. 4.23 PDF of reconstruction errors in the particle centres for both Tomo-PIV and LF-PIV analysis of synthetic particle images. **a** Tomo-PIV, 0.05PPP, **b** Tomo-PIV, 0.1PPP, **c** LF-PIV, low LTPR and **d** LF-PIV, high LTPR

accuracies in all three directions can be seen to increase correspondingly. In particular, a PMR = 14 value appears to strike a good compromise between PMR and MLA resolution. Moving on to the results for Tomo-PIV, it is within expectations to see that the measurement errors reduce when more cameras were used (Elsinga et al. 2006; Atkinson and Soria 2009), particularly in the *y* direction. This is because the *y* direction is parallel to all the cameras and more cameras will serve to improve the reconstruction quality along that direction. Also, in contrast to LF-PIV where ghost particles do not pose a problem, they are presented here and could be behind the observation that a higher particle density of 0.1PPP produces higher errors than 0.05PPP (Elsinga et al. 2006; Scarano 2013). The key takeaway from the above is that LF-PIV is able to produce comparable results when compared to Tomo-PIV, at least based on the DNS results here, even for a relatively low pixel resolution ratio test case such as the one associated with LTPR = 3.13, 4.17, 6.25 at PMR = 7. In

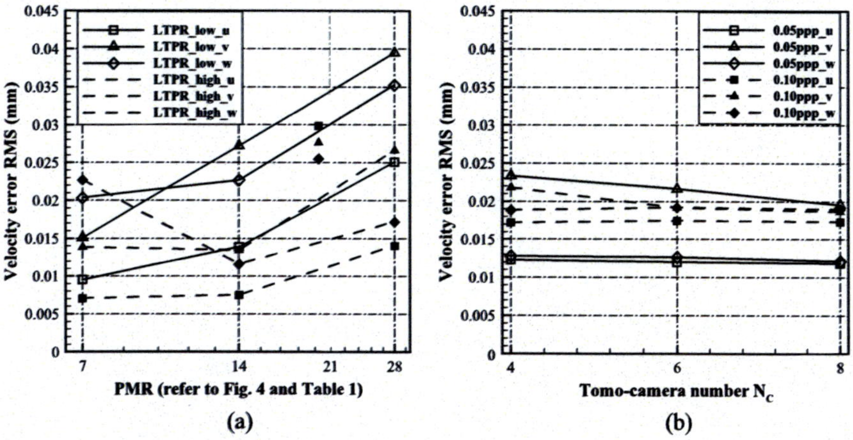

Fig. 4.24 Velocity error RMS when compared with DNS results. **a** LF-PIV for the low (PTPR = 3.13, 4.17, 6.25) and high resolution (PTPR = 12.5, 16.67, 25) light-field cameras **b** Tomo-PIV (pixel size = 0.075 mm)

fact, the present analysis shows that LF-PIV has the potential to produce comparable or better accuracy levels than Tomo-PIV for the same field-of-view, especially when camera and MLA resolution continue to improve over time.

Last but not least, actual experiments were conducted to move the comparison beyond simply the use of synthetic particle images. The experimental setup used to experimentally validate LF-PIV earlier was used and hence its physical details will not be described here again. Instead, Fig. 4.25 shows the physical camera arrangements for LF-PIV and Tomo-PIV configurations. Particle densities used for Tomo-PIV and LF-PIV tests were set at 0.062PPP and 0.06PPM respectively. The light-field camera was the one described earlier in the chapter, whilst four 4-megapixel Imperx B2014 PIV cameras coupled with Micro-Nikkor 85 mm Scheimpflug lenses were used for Tomo-PIV. Effective resolutions for LF-PIV and Tomo-PIV were determined to be 6600×4400 pixels and 480×320 pixels respectively, which represents a LTPR = 47.27 scenario. General reconstruction methodologies for LF-PIV and Tomo-PIV were similar to the ones used during synthetic particle image analysis, though 400 and 40 iterations were used for the former and latter approaches respectively. Reconstruction domains for LF-PIV and Tomo-PIV were $3300 \times 2200 \times 182$ voxels and $480 \times 320 \times 130$ voxels respectively as well. Velocity fields were obtained by subjecting the reconstructed 3D particle images to multi-grid correlations with 75% overlapping ratio. Initial and final interrogation windows were $320 \times 320 \times 64$ voxels and $160 \times 160 \times 32$ voxels for LF-PIV, whilst they were $64 \times 64 \times 64$ voxels and $32 \times 32 \times 32$ voxels for Tomo-PIV. Details of these post-processing parameters are presented in Table 2 of Shi et al. (2018). Similar to the earlier validation exercise, the goal here was to capture the vortex roll-ups along the jet shear layer, and a side-by-side comparison of how well both approaches capture the same instantaneous vortex roll-up flow field can be seen in Fig. 4.26. It can be deduced

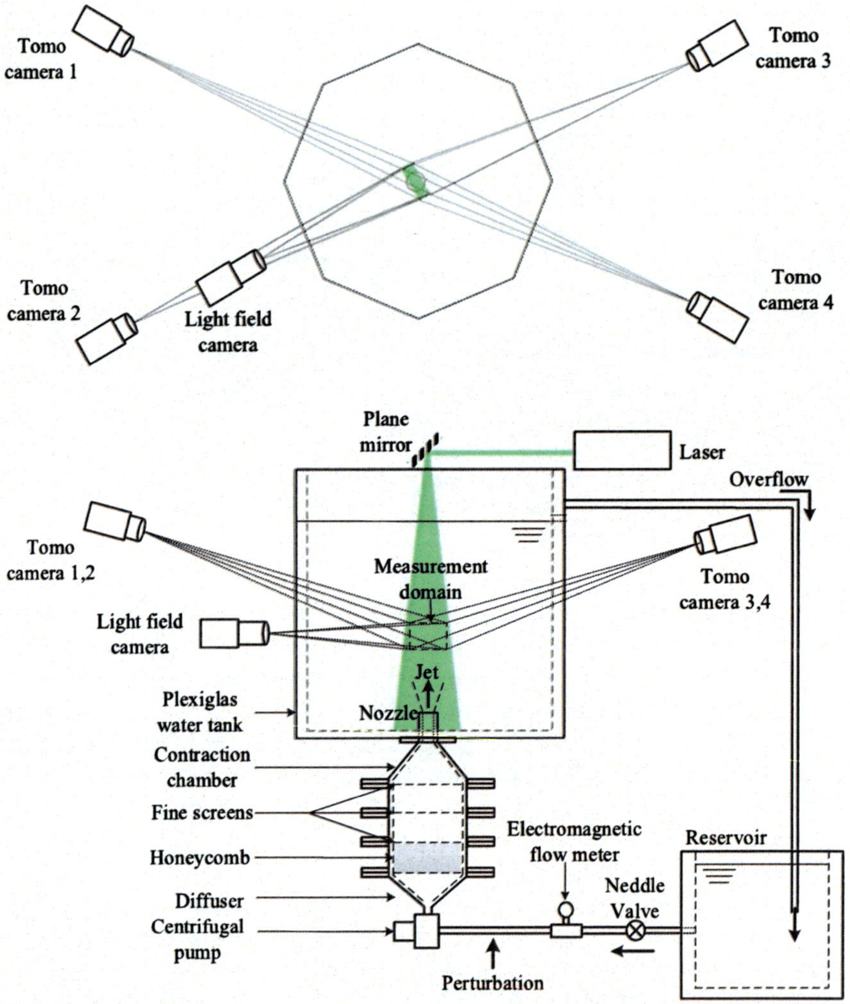

Fig. 4.25 Schematics of the experimental setup and physical camera arrangements used for LF-PIV and Tomo-PIV approaches

that the vortex roll-up captured by the two different approaches are similar and with a correlation level of 0.94 between the two velocity fields, one can appreciate that the outcome of LF-PIV approach is very comparable with Tomo-PIV, at least for the present jet flow experiments.

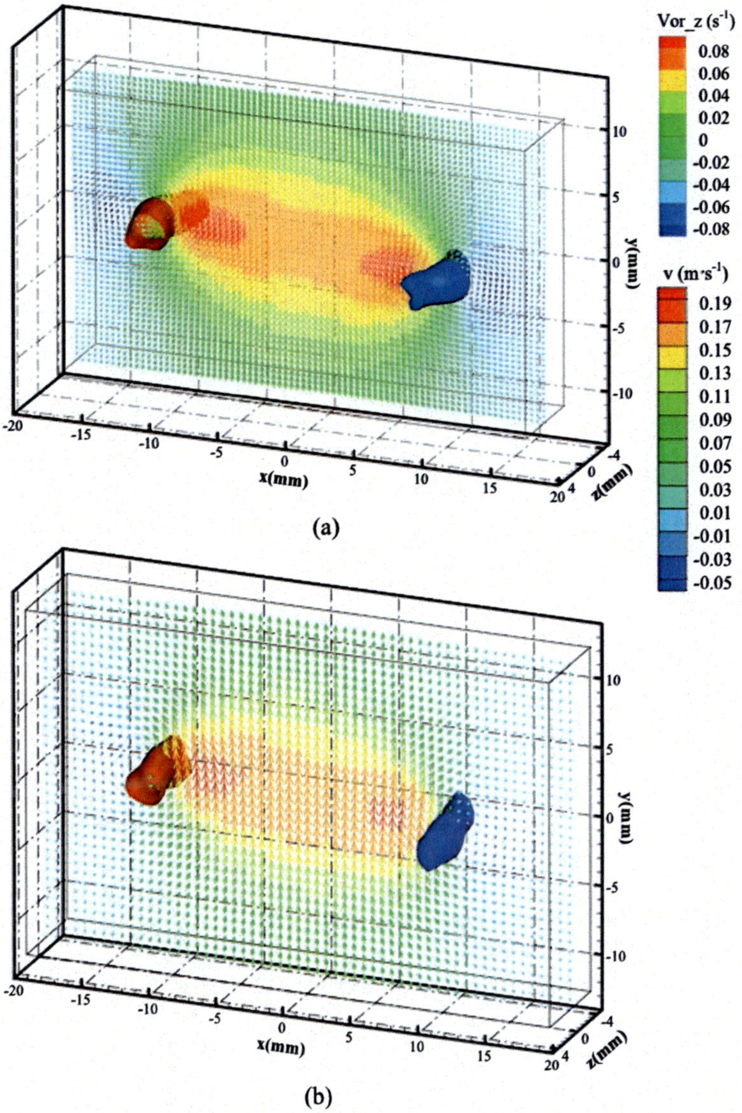

Fig. 4.26 Side-by-side comparison of the instantaneous velocity and vorticity fields obtained by **a** LF-PIV and **b** Tomo-PIV

Dual-Camera LF-PIV

One of the issues encountered when developing a single-camera LF-PIV technique is the higher uncertainties in the depth direction when small viewing angles are used, as what we had seen previously. Using synthetic light-field images of one single PIV

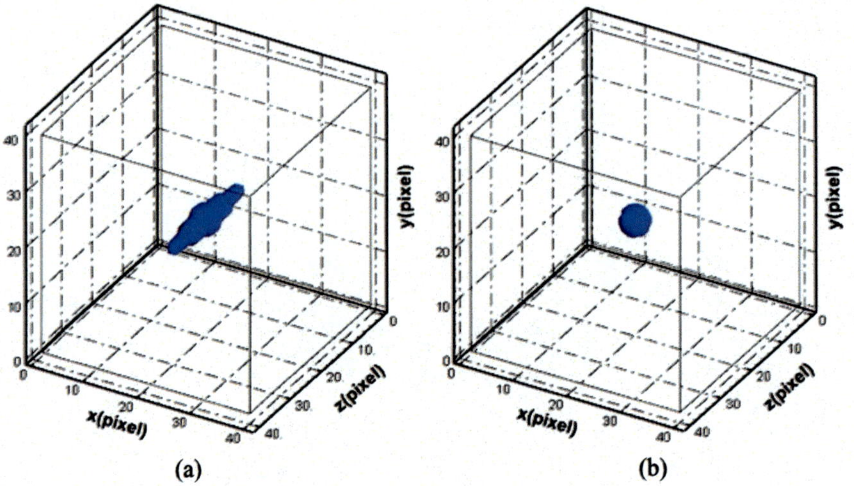

Fig. 4.27 Reconstruction results for a PIV particle generated from a synthetic image based on **a** single light-field camera approach and **b** dual light-field camera approach

particle as an example, the reconstruction result for a single light-field camera can be appreciated in Fig. 4.27a, where the reconstruction result shows a reconstructed particle that is stretched in the depth direction. Whilst the centroid of the reconstructed particle can still be successfully identified by the in-house post-processing algorithm, note that an idealised round seeding particle was assumed in generating the synthetic light-field particle images. In most actual PIV measurements, such an assumption is very unlikely to be realised, on top of other considerations such as size variations, particle rotations and other factors. As such, a study was conducted by Mei et al. (2019) to see how much benefit will be by adding one more light-field camera be when it comes down to the accuracy levels of particle reconstruction and tracking for PIV purposes.

In the study, using the same synthetic light-field images for one single idealised PIV particle but reconstructed with two light-field cameras instead of just one, the reconstructed particle is now much closer to its idealised geometry, as can be seen in Fig. 4.27b. More importantly, the centroid of the reconstructed particle can now be ascertained with greater accuracy, which would in turn translate to higher accuracy levels for LF-PIV measurements. These initial results gave confidence to the possibility of achieving a significant increase in LF-PIV accuracy levels by simply using another light-field camera for flow scenarios. Hence, additional tests were carried out to further quantify the impact of using dual light-field camera approach, as compared to single light-field camera. In principle, implementation of two light-field cameras for dual LF-PIV 3D measurements of a flow scenario will be relatively straight forward and resemble the procedures shown in Fig. 4.28. Briefly speaking, two light-field cameras will be viewing the flow scenario from two different perspectives, with each camera recording its own light-field particle images and synchronised

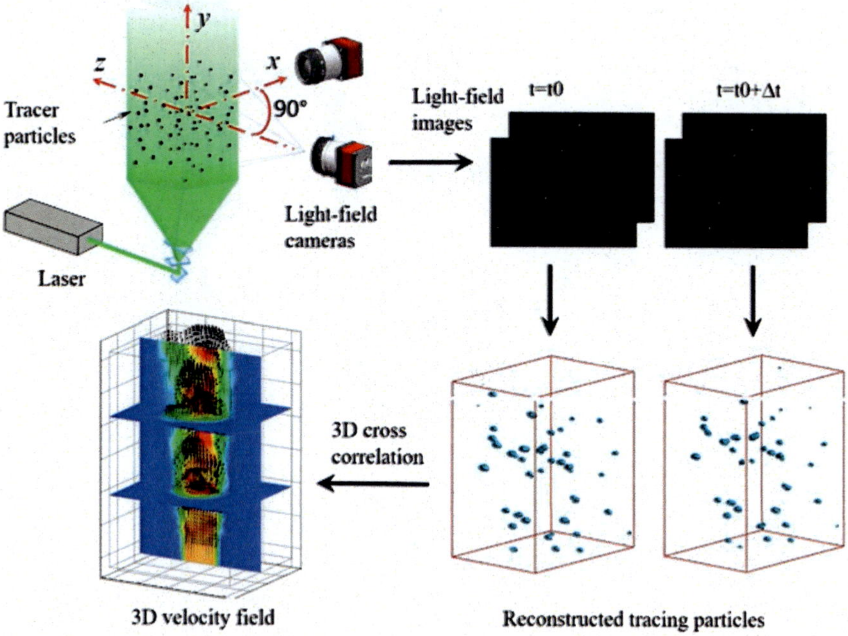

Fig. 4.28 General schematics of procedures associated with dual LF-PIV measurements of a flow scenario

with the other camera. The reconstruction process will take into account the separation angle between the two light-field cameras, before processing the two image sets and arriving at the final combined locations of the 3D particle images. With an additional light-field camera incorporated, the reconstruction process of a dual LF-PIV setup is schematically represented using ray-tracing in Fig. 4.29, from the voxels to the main lens, MLA and eventually the imaging sensor (CCD/CMOS). The volumetric calibration and reconstruction process can be summarised via the flowchart presented in Fig. 4.30 and interested readers can refer to Mei et al. (2019) for more details in the reconstruction algorithms and other details.

Before a dual camera LF-PIV approach can be implemented however, the basic principles will need to be validated even if things seem straight forward. And just with the case of a single camera LF-PIV, the effects of experimental parameters such as camera separation angle, particle density and others will have to be understood first, before dual camera LF-PIV can be used for actual experimental flow validations. This will be elaborated in greater detail in the next section.

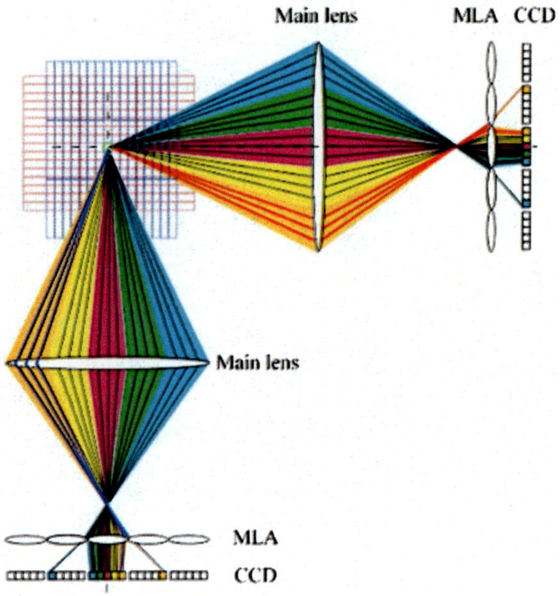

Fig. 4.29 Schematics of ray-tracing from the voxels in the measurement volume to the imaging sensor via the main lens and MLA for the two light-field cameras, which will be used for the two-camera reconstruction approach

Initial Validation of Dual Camera Principles

To begin with, initial validations using real-world images of multiple tiny spherical glass beads held up by very slender syringe needles and located at various 3D locations were taken by a single light-field camera, as shown schematically in Fig. 4.31a. To simulate the use of two light-field cameras, a rotating platform was used to rotate the glass bead setup through discrete angular locations during the image capturing process, as shown in Fig. 4.31b. Samples of the light-field images captured by the light-field camera at 0° and 90° are shown in Fig. 4.31c and d to better show the differences in the images between the two orientations. Through this way, the locations and geometries of the glass beads can be reconstructed by considering a single camera approach or dual camera approach with different angular differences between them. This allowed better understanding of the impact of and optimal angular locations in a dual light-field camera approach, even before an actual second light-field camera is used.

To understand how the implementation of a second light-field camera could lead to better reconstruction of the imaged glass beads in the 3D space, an example based on light-field images taken at 0° and 90° angular locations will be briefly described here. Figure 4.32a and b shows the reconstructed results for each of the light-field images taken at these two angular locations. Note that both of them exhibit elongated geometries in the reconstructed results—the one at 0° angular location shows elongations along the Z-direction, whilst the one at 90° angular location shows elongations along the X-direction. Taken individually, each result demonstrates the higher uncertainty levels expected from the use of a single light-field camera. However, if the

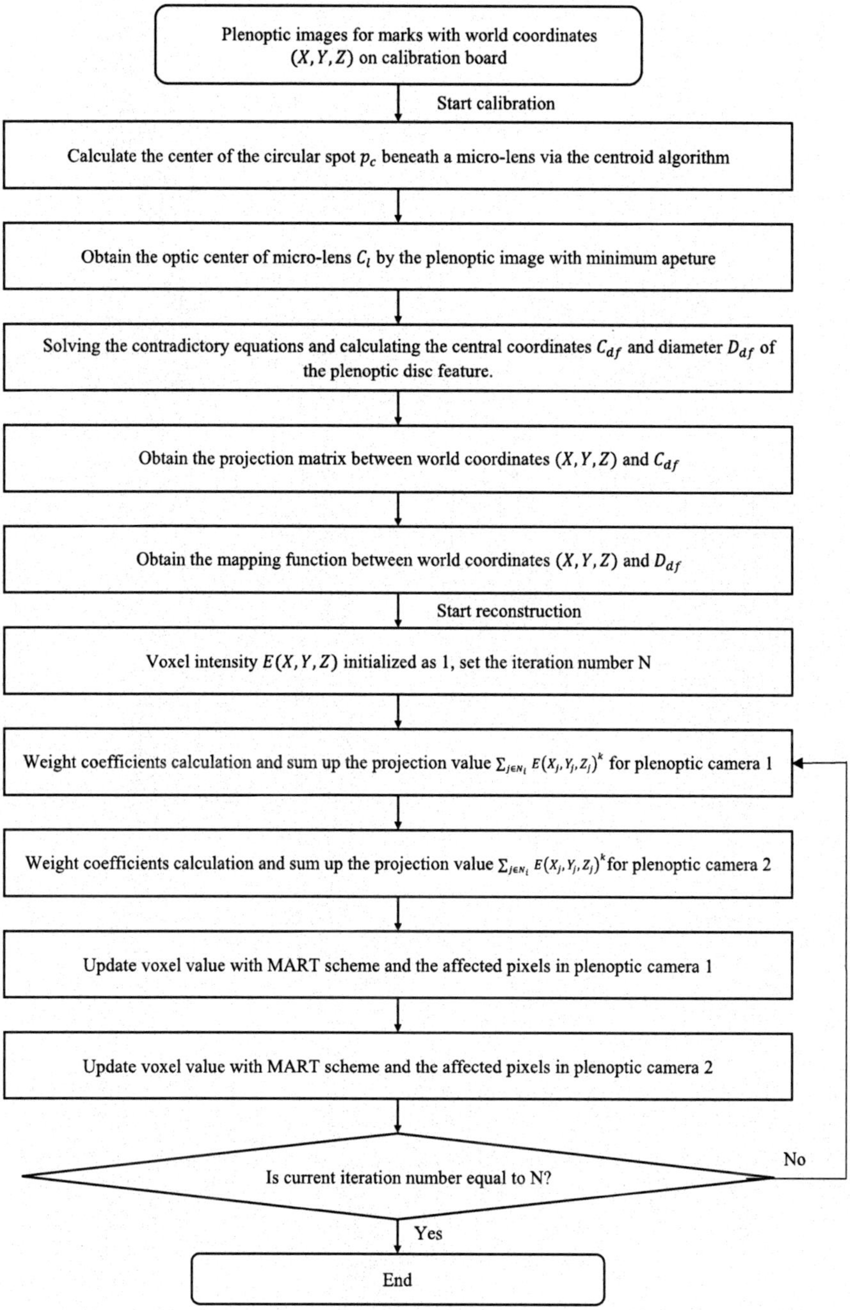

Fig. 4.30 Flowchart that outlines the volumetric calibration and particle reconstruction processes for dual camera LF-PIV

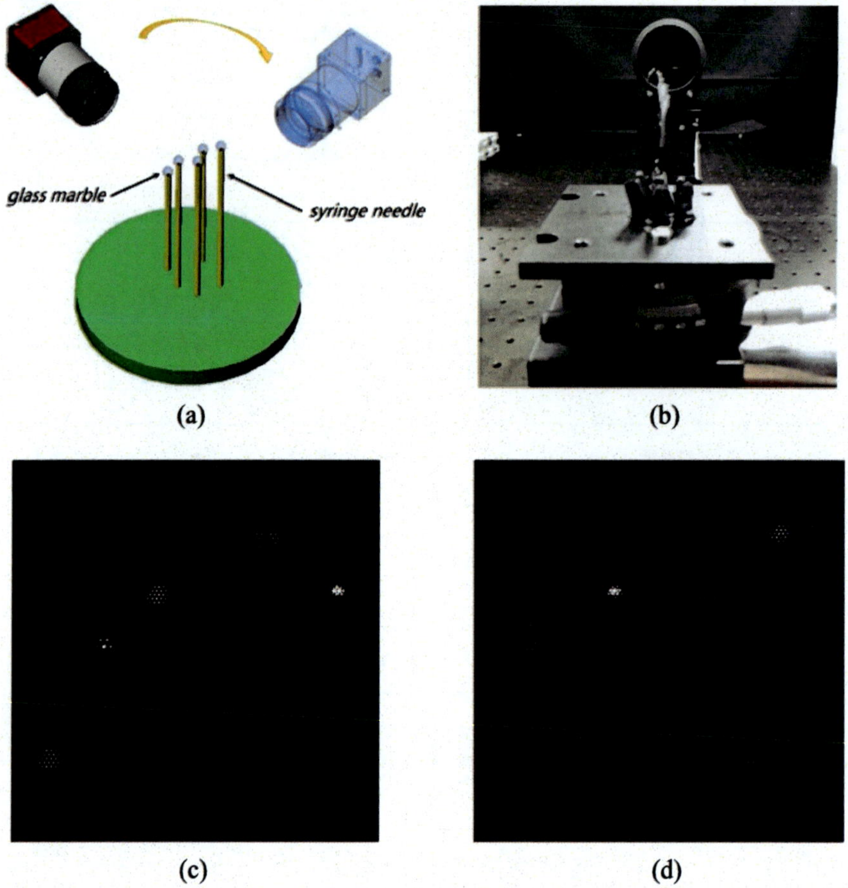

Fig. 4.31 **a** Schematics of the glass bead 3D positioning setup; **b** photo of the light-field camera and glass bead setup; **c** and **d** raw light-field images captured by the camera at 0° and 90° angular locations

results from both angular locations are combined, significantly better reconstruction results without discernible elongations in any direction can be achieved, as shown in Fig. 4.32c. With this successful demonstration, dual camera LF-PIV approach will now be studied for the effects of particle density and separate angle based on synthetic light-field particle images.

Effects of Particle Density and Separation Angle

It has been established in earlier studies by the authors that a particle density of 0.5PPM strikes a good balance between reconstruction quality and sample window

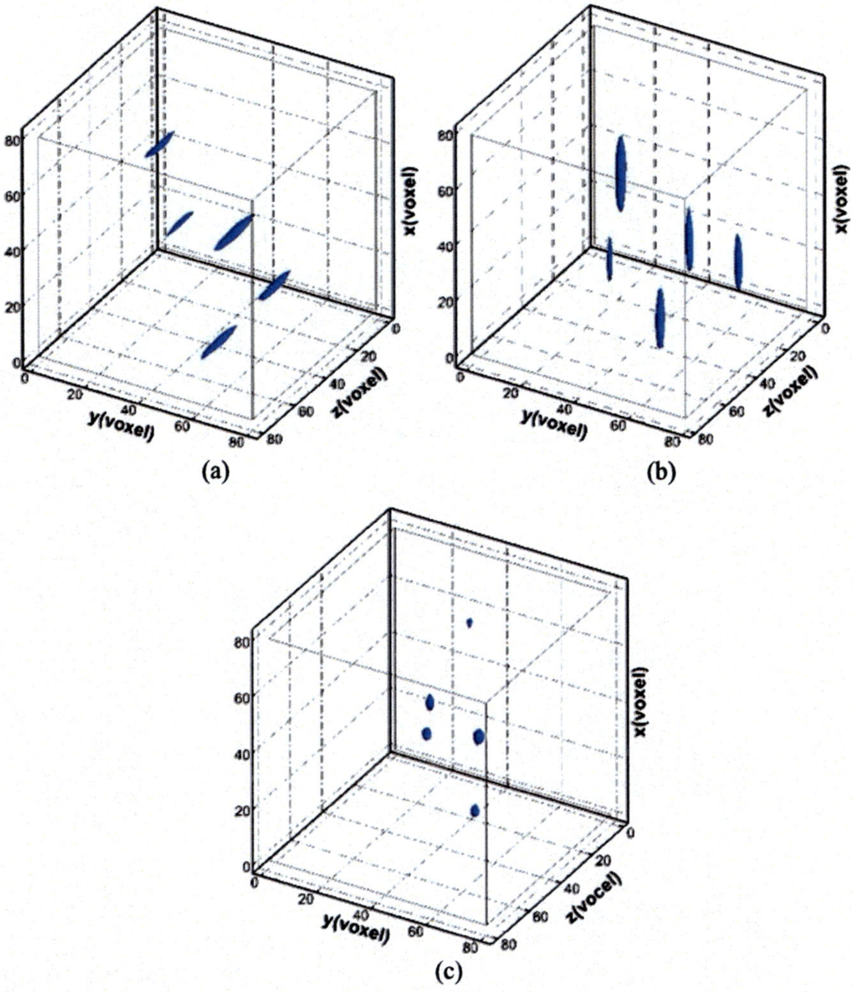

Fig. 4.32 Reconstructed results for light-field images taken at **a** 0° and **b** 90°, as well as based on both orientations for dual light-field camera approach

size. However, this is unclear for a dual camera LF-PIV setup, so a series of tests were conducted to evaluate the impact of particle density on dual camera LF-PIV approach more thoroughly. In this case, 0.1, 0.5 and 1.0PPM will be used. Synthetic light-field images with these particle densities were generated and reconstructed, before the reconstructed particle intensities in the volume were compared with the original one that was used to generate the synthetic light-field images. This was done by using reconstruction quality Q_{Recon}, that was described and used for single camera LF-PIV earlier. Figure 4.33 shows the variations in the reconstruction quality with iteration number under different particle density and light-field camera configurations, where several interesting observations can be made. Firstly, the reconstruction quality is

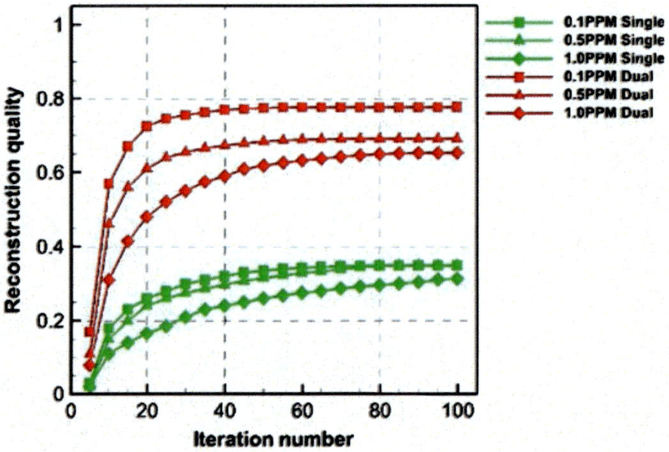

Fig. 4.33 Variations in the reconstruction quality with iteration number under different particle density and light-field camera configurations

significantly better when a dual camera setup is used for the same iteration number. Secondly, a lower particle density level produces a higher reconstruction quality, similar to what had been observed previously for single camera LF-PIV. Thirdly, the convergence of the reconstruction quality is more rapid for the dual camera setup. And lastly, a dual camera setup is able to handle higher particle density levels much better than a single camera setup. This can be attributed to the fact that two light-field cameras provide wider perspectives of the particles than a single light-field camera, which leads to less elongated reconstructed particles and improve reconstruction quality.

Next, synthetic light-field particle images based on different viewing orientations of two different light-field cameras of the same volume were constructed at particle densities of 0.5 and 1PPM. Note that the two different viewing orientations were characterised as the separation angle between them, where 0° and 180° refers to the two cameras having the same and directly opposing viewing orientations. The former configuration would have reverted back to a single-camera-based LF-PIV approach. Figure 4.34 shows how the reconstruction quality varies between 0° and 180° at 10° intervals, where it reaches a peak level at 90° separation angle. Hence, it is expected that this separation angle would be optimal for subsequent experimental validations. Additionally, halving the particle density from 1PPM to 0.5PPM leads to a small increase in reconstruction quality as well. An optimal separation angle of 90° can be understood if one considers the effects upon the spatial resolution distribution along the out-of-plane direction (i.e. z direction for each camera) as the separation angle varies. Two light points are considered to be distinguishable when their light rays reach two separate groups of microlenses, whereas they will be less distinguishable if these light rays are recorded by the same group of microlenses.

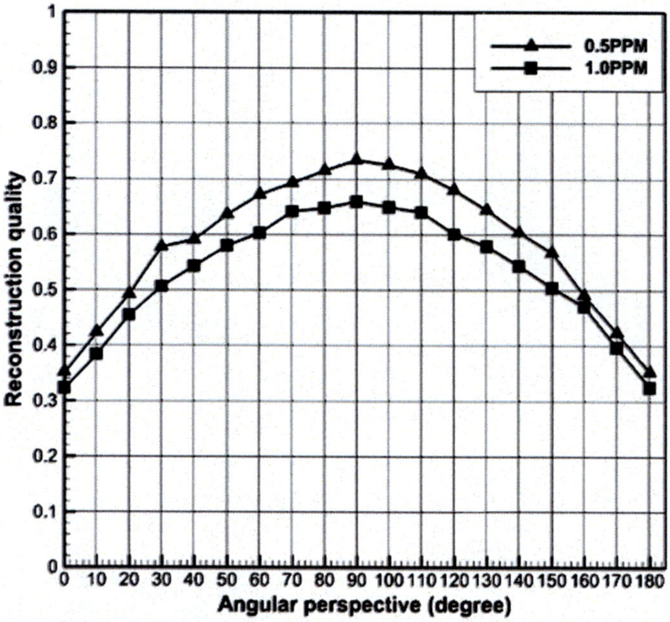

Fig. 4.34 Comparison of reconstruction quality due to variations in the angular perspective or separation angle between the cameras in a dual camera setup

Hence, a 90° separation angle leads to the most optimal configuration that will allow the best differentiation between the two different light points.

Recall that the objective of having a dual camera LF-PIV approach is to mitigate particle elongations caused by a lower resolution in the depth direction if only a single camera were to be used. Now, the resolution in the depth direction for a dual camera setup will be examined for various separation angles and Fig. 4.35 shows the results for 0° to 90° separation angles at 30° intervals. In the plots, z direction is the depth direction, where in this case for two cameras will be the angular bisector. The origin will also be located where the two optical axes intersect. Note also that the legend is the depth resolution in millimetres. For 0° separation angle, where the optical axes of the two cameras are aligned, the situation resembles that of a single camera configuration and is not expected to demonstrate better improvement beyond that. This can be seen in Fig. 4.35a where resolution is about 0.55 mm along a strip along the x–z plane and $y = 0$ direction. As the separation angle increases gradually to 90°, two effects can be observed. Firstly, the region with poorer resolution reduces in size and secondly, the resolution improves to about 0.1 mm by the time the separation angle reaches 90°. These findings agree well with the earlier notion that 90° is the optimal separation angle.

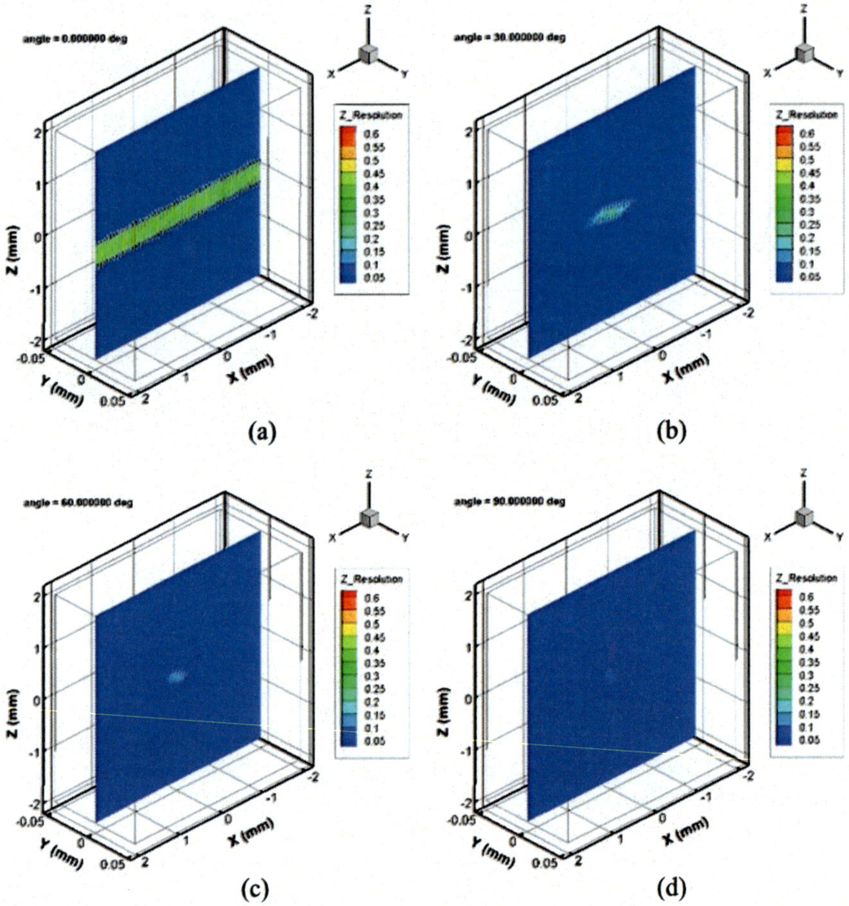

Fig. 4.35 Resolution in the depth direction for dual camera separation angles of **a** 0°, **b** 30°, **c** 60° and 90°

Validations with Simulation and Experiment

Just as with the case for single camera LF-PIV, validations with increasingly more realistic flow scenarios were carried out and here, dual camera LF-PIV approach will be tested against DNS results of a Re = 2500 incompressible jet flow. Data from the simulation results were used to create corresponding synthetic light-field images for a volume that was located at one jet diameter (D) above the nozzle exit, where the volume has a size of $0.66D \times 0.66D \times 0.66D$. Particle density used for the synthetic light-field images was 1PPM and these particles were randomly dispersed within the first frame. Subsequently, the particles will undergo displacements based on the DNS 3D velocity results and 2 ms time intervals, before they were used to generate the synthetic light-field images. Both single and dual camera LF-PIV approaches were then used to process for the final 3D velocity fields and compare with the DNS results. For the former, the reconstruction volume comprised of 800 × 800 × 267 voxels, whilst the pixel-to-voxel ratio is 3:1 in the x and y directions but 10:1 in the z direction. As for the latter, the reconstruction volume was 800 × 800 × 800 voxels, whilst a pixel-to-voxel ratio of 3:1 was used for all three directions. The voxel number and pixel-to-voxel ratio was different for the dual camera LF-PIV approach, since the computational time was essentially double that of a single camera LF-PIV. It should also be mentioned that GPU acceleration through the use of Nvidia CUDA codes was adopted to reduce computational times. To obtain the final velocity fields, multi-grid cross-correlation and filters similar to those used earlier on were used here as well.

Figure 4.36 shows a comparison between the velocity fields taken along different planes from the DNS, single camera and dual camera LF-PIV results. At first glance, the agreement between the LF-PIV results and DNS data appeared to be excellent, where jet ring vortices were captured regardless of whether a single or dual camera approach was used. Whilst little visual differences can be observed in the plots

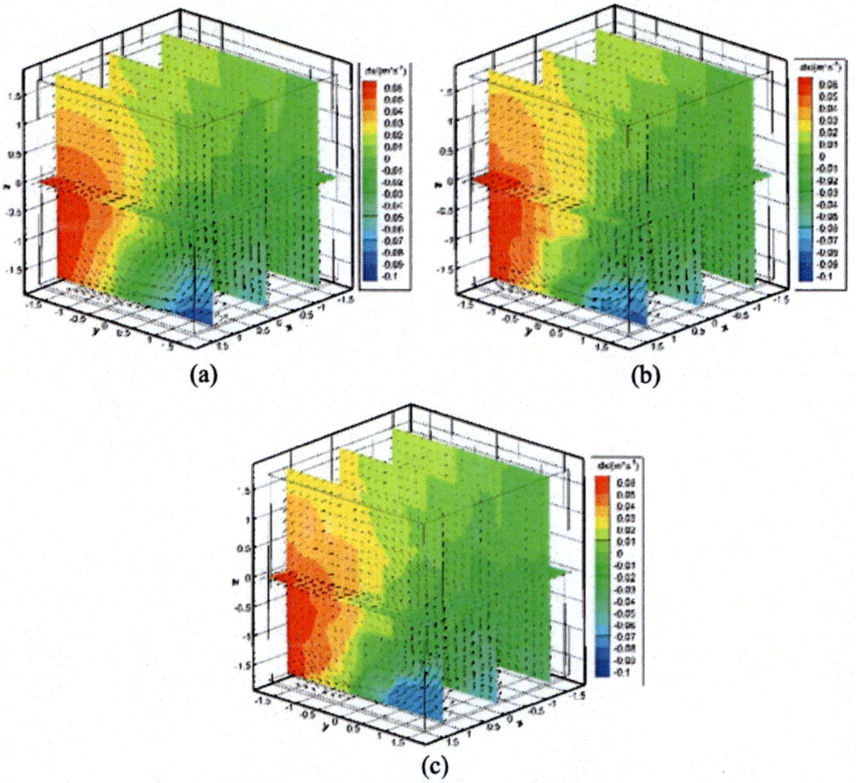

Fig. 4.36 Flow fields plotted from the **a** original DNS data, **b** single camera LF-PIV post-processed results and **c** dual camera LF-PIV post-processed results

presented in Fig. 4.36, a more systematic comparison between the cumulative distribution functions of the measurement errors along the x, y and z directions will paint a better picture. Note that x–y plane is normal to the first camera optical axis, whilst z direction is the optical axis direction. Based on the results shown in Fig. 4.37, it can be discerned that the accuracy levels along the x–y plane do not really depend upon whether single or dual camera LF-PIV approach was used. In contrast, the accuracy level in the z direction clearly benefits from dual camera LF-PIV approach and in fact approaching to that of the $x - y$ plane. As such, this analysis demonstrates the capability of dual camera LF-PIV approach to achieve more accurate measurement results in the z direction (i.e. depth direction) over a single camera based approach. After testing out single and dual camera LF-PIV approaches against simulated DNS data, the next logical step would involve the use of dual camera setup on actual experiments to confirm what had been learned thus far. For this experimental test, a

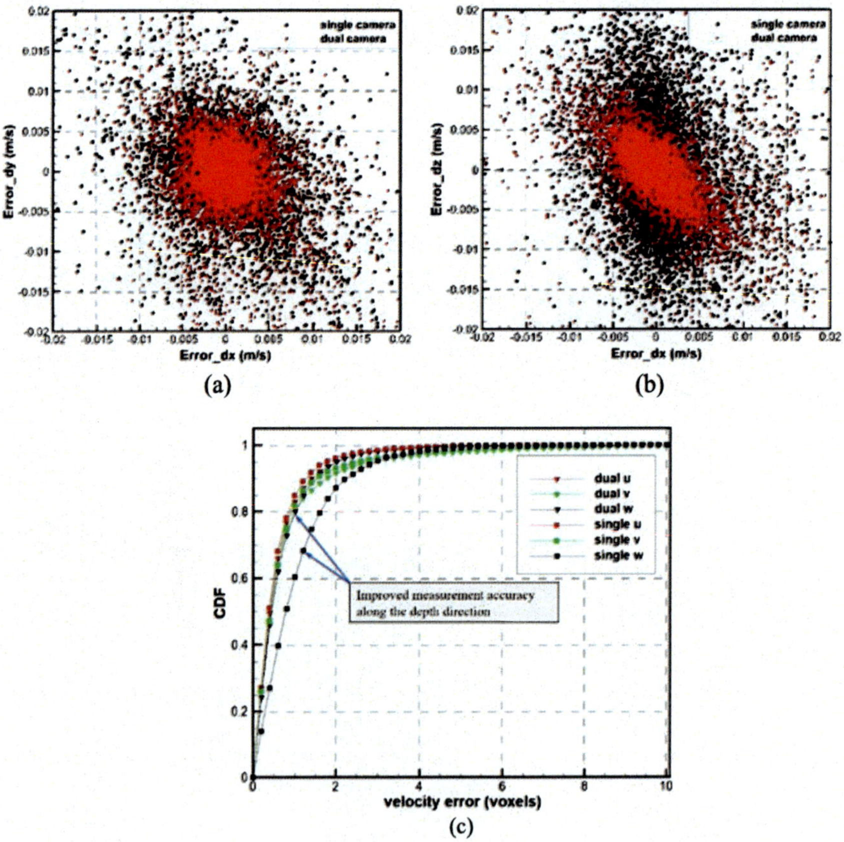

Fig. 4.37 Comparisons between the cumulative distribution functions of velocity measurement errors in x, y and z directions

canonical and reproducible flow scenario would be preferred and, in this case, well-understood laminar circular vortex rings were used to test out an actual dual camera LF-PIV setup, workflow and analysis outcomes.

Figure 4.38 shows the water-based experimental setup used to generate the discrete vortex-rings, where a "cylindrical slug" approach was utilised. Similar setups had been used in earlier vortex-ring studies (New et al. 2016, 2020; New and Zang 2017) and hence, they will be briefly described here. A high-torque stepper motor programmed on a workstation was used to drive a piston impulsively through a stainless-steel tube and push water out of a $D = 20$ mm circular nozzle into a quiescent environment. A stroke length of $L/D = 1.5$ was used such that it was shorter

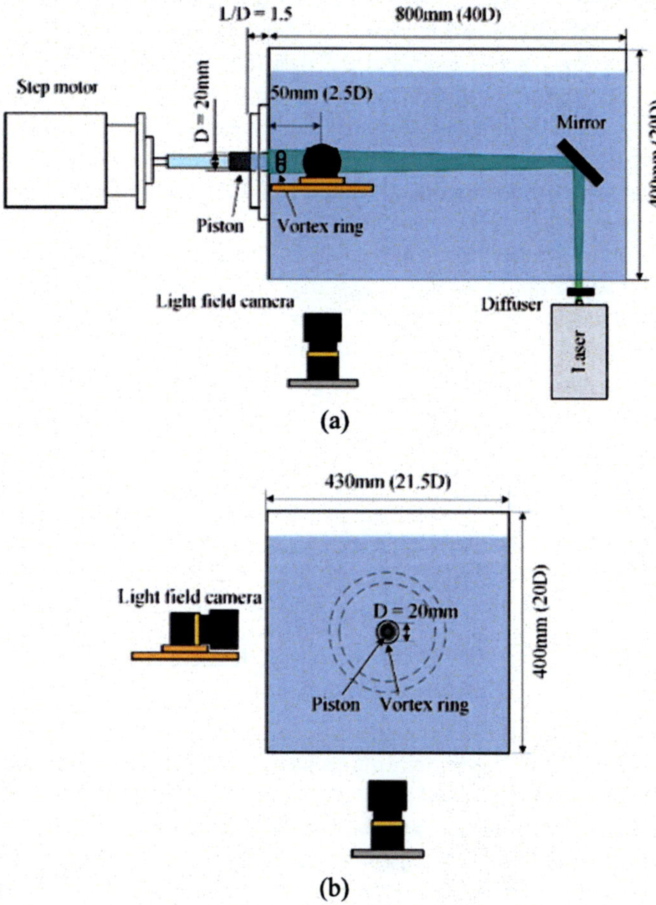

Fig. 4.38 Schematics of the vortex-ring experimental setup used for the testing of dual camera LF-PIV approach, where **a** shows the side-view and placement of the first light-field camera by the side of the water tank, whilst **b** shows that the second light-field camera is below the water tank. These camera placements ensured that an optimal separation angle of 90° was used

than the formation length and prevented any trailing jet to be formed aft of the vortex-rings. Trapezoidal velocity profiles with sharp acceleration and deceleration stages were used and the resulting vortex-rings have an approximate Reynolds number of 2000. Two light-field cameras with specifications similar to those used in earlier studies were positioned outside of the water tank, such that they were orthogonal to each other as shown in figure. 200 mm Micro-Nikkor lenses were mounted onto both light-field cameras and set to a consistent f-number of 4.0, with the time interval between the image-pairs set to 2 ms. Volumetric calibration was carried out prior to the experiments using a matt black board with white dots spaced at 3 mm intervals in both vertical and horizontal directions. The calibration board was translated within the intended measurement volume through the use of a high-precision, motorised translation stage with a 0.1 μm resolution. As the calibration effectiveness increases with the number of calibration planes used, a total of 51 calibration images were taken by both light-field cameras at 0.5 mm intervals. 50 μm polyamide seeding particles were uniformly distributed in the water within the entire water tank and vortex-ring producing tube/nozzle and particle density was estimated to be approximated 0.431PPM, which was close to the optimal 0.5PPM mentioned earlier. These seeding particles were subsequently illuminated by a 500 mJ/pulse, double-pulsed Nd:YAG laser during the experiments.

To maximise the measurement resolution, the measurement volume was set to a relatively small 32 mm × 24 mm × 32 mm physical size, which was just about sufficiently large to contain the entire vortex-ring. On the other hand, the computational load associated with post-processing for high-resolution results had to be taken into consideration and a pixel-to-voxel ratio of 3:1 was used, instead of 2:1 ratio. This resulted in a more manageable 1904 × 1404 × 1904 voxels, as opposed to 2856 × 2106 × 2856 voxels if 2:1 ratio was to be used. The latter voxel density would see each reconstructed volume ballooning to more than 64 GB in size and hence, not feasible at this point. Last but not least, in-house DRT-MART-based post-processing was used to arrive at the final single and dual camera LF-PIV results that are shown in Fig. 4.39. From the figure, it can be observed from the comparison (especially between Fig. 4.39a(i) and b(i)) that the dual camera approach was able to capture the vortex-core and its swirling nature much better than its single camera counterpart. Essentially, this means that a dual camera LF-PIV approach leads to significantly lower measurement errors, which agrees well with the earlier analysis. More accurate velocity field measurements as demonstrated in Fig. 4.39b(i) also means that the isosurfaces of the vortex-ring as based on vorticity magnitudes, whilst not quite perfect, are better mapped out under a dual camera approach as well, when one compares between Fig. 4.39a(ii) and b(ii), as well as between Fig. 4.39a(iii) and b(iii). To further quantify the measurement accuracy levels, probability density functions of the divergence error for the measured velocity vectors between single and dual camera LF-PIV approaches here with and without reconstruction post-processing were determined and presented in Fig. 4.40. The idea of using divergence error as a measure of accuracy stemmed from the consideration of the continuity equation associated with incompressible fluids, with the divergence error defined here as

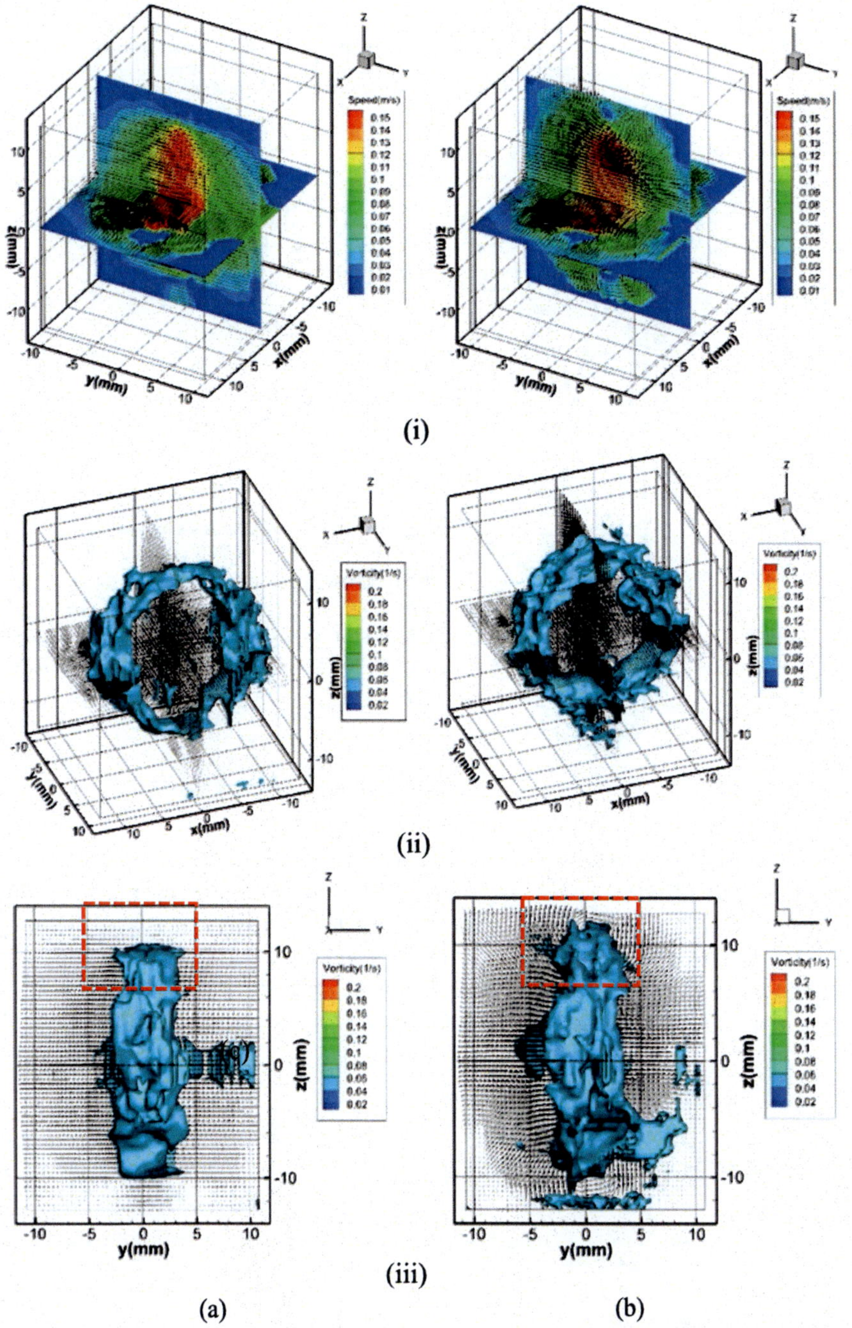

Fig. 4.39 Comparisons between instantaneous flow fields resulting from **a** single camera and **b** dual camera LF-PIV approaches

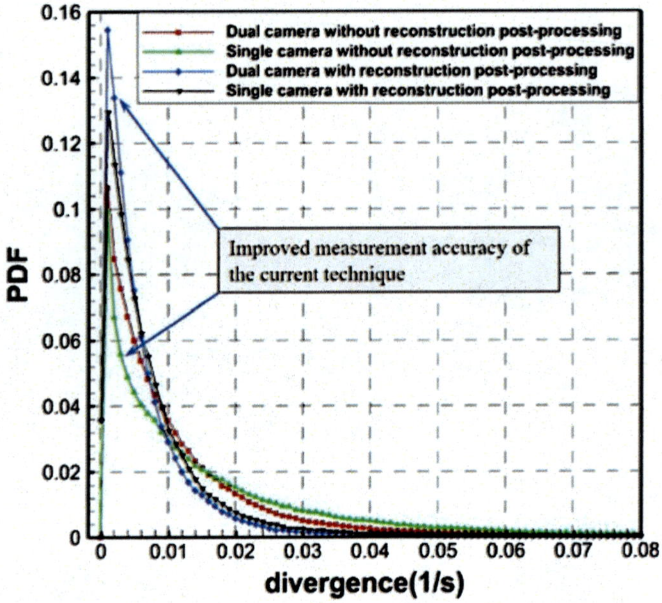

Fig. 4.40 Comparison of probability density functions (PDF) of the measured velocity vectors between single and dual camera LF-PIV approaches with and without reconstruction post-processing

$$\text{Error}_{\text{divergence}} = \left| \frac{\partial u}{\partial x} + \frac{\partial v}{\partial y} + \frac{\partial w}{\partial z} \right| \tag{4.6}$$

From the results depicted in Fig. 4.40, it can be discerned that most of the results associated with no reconstruction post-processing have divergence errors of no more than 0.04 s^{-1}, whilst those with reconstruction post-processing have divergence errors of typically less than 0.02 s^{-1}. This represents a significant drop in the error level and further reinforces the notion that a dual camera setup goes a long way towards more accurate measurement results than what a single camera setup is able to provide.

References

Adelson EH, Wang JYA (1992) Single lens stereo with a plenoptic camera. IEEE Trans Pattern Anal Mach Intell 14:99–106

Atkinson C, Soria J (2009) An efficient simultaneous reconstruction technique for tomographic particle image velocimetry. Exp Fluids 47:553–568

Deem EA, Zhang Y, Cattafesta LN et al (2016) On the resolution of plenoptic PIV. Meas Sci Technol 27:84003

Elsinga GE, Scarano F, Wieneke B, Van Oudheusden BW (2006) Tomographic particle image velocimetry. Exp Fluids 41:933–947. https://doi.org/10.1007/s00348-006-0212-z
Fahringer TW, Lynch KP, Thurow BS (2015) Volumetric particle image velocimetry with a single plenoptic camera. Meas Sci Technol. https://doi.org/10.1088/0957-0233/26/11/115201
Fahringer TW, Thurow BS (2012) Tomographic reconstruction of a 3-D flow field using a plenoptic camera. In: 42nd AIAA fluid dynamics conference and exhibit 2012
Fahringer TW, Thurow BS (2013) The effect of grid resolution on the accuracy of tomographic reconstruction using a plenoptic camera. In: 51st AIAA aerospace sciences meeting including the new horizons forum and aerospace exposition 2013
Fahringer TW, Thurow BS (2015) Comparing volumetric reconstruction algorithms for plenoptic-PIV. In: 53rd AIAA aerospace sciences meeting
Fahringer TW, Thurow BS (2014) 3d particle position reconstruction accuracy in plenoptic piv. In: 52nd aerospace sciences meeting, p 398
Georgiev T, Intwala C (2006) Light field camera design for integral view photography. Adobe Tech Rep 1:13
Hall EM, Fahringer TW, Guildenbecher DR, Thurow BS (2018) Volumetric calibration of a plenoptic camera. Appl Opt 57:914–923
Hall EM, Guildenbecher DR, Thurow BS (2019) Development and uncertainty characterization of 3D particle location from perspective shifted plenoptic images. Opt Express 27:7997–8010
Klemkowsky JN, Fahringer TW, Clifford CJ et al (2017) Plenoptic background oriented Schlieren imaging. Meas Sci Technol 28:95404
Long J, New TH (2015) A DPIV study on the effects of separation distance upon the vortical behaviour of jet–cylinder impingements. Exp Fluids 56:1–21
Long J, New TH (2016) Vortex dynamics and wall shear stress behaviour associated with an elliptic jet impinging upon a flat plate. Exp Fluids 57:1–18
Long J, New TH (2019) Vortical structures and behaviour of an elliptic jet impinging upon a convex cylinder. Exp Therm Fluid Sci 100:292–310
Lumsdaine A, Georgiev T (2009) The focused plenoptic camera. In: 2009 IEEE international conference on computational photography (ICCP). IEEE, pp 1–8
Lynch K, Fahringer T, Thurow B (2012) Three-dimensional particle image velocimetry using a plenoptic camera. In: 50th AIAA aerospace sciences meeting including the new horizons forum and aerospace exposition
Lynch KP, Thurowy BS (2011) Preliminary development of a 3-D, 3-C PIV technique using light field imaging. In: 41st AIAA fluid dynamics conference and exhibit
Mei D, Ding J, Shi S et al (2019) High resolution volumetric dual-camera light-field PIV. Exp Fluids. https://doi.org/10.1007/s00348-019-2781-7
New TH, Long J, Zang B, Shi S (2020) Collision of vortex rings upon V-walls. J Fluid Mech 899
New TH, Shi S, Zang B (2016) Some observations on vortex-ring collisions upon inclined surfaces. Exp Fluids 57:1–18
New TH, Tsai HM (2007) Experimental investigations on indeterminate-origin V- and A-notched jets. AIAA J 45:828–839
New TH, Tsovolos D (2012) On the flow characteristics of minor-plane inclined elliptic jets. Exp Therm Fluid Sci 38:94–106
New TH, Zang B (2017) Head-on collisions of vortex rings upon round cylinders. J Fluid Mech 833:648–676
Ng R (2006) Digital light field photography. Stanford University
Roberts WA, Thurow BS (2017) Correlation-based depth estimation with a plenoptic camera. AIAA J 55:435–445
Scarano F (2013) Tomographic PIV: principles and practice. Meas Sci Technol 24. https://doi.org/10.1088/0957-0233/24/1/012001
Shi S, Ding J, Atkinson C et al (2018) A detailed comparison of single-camera light-field PIV and tomographic PIV. Exp Fluids. https://doi.org/10.1007/s00348-018-2500-9

Shi S, Ding J, New TH, Soria J (2017) Light-field camera-based 3D volumetric particle image velocimetry with dense ray tracing reconstruction technique. Exp Fluids. https://doi.org/10.1007/s00348-017-2365-3

Shi S, New TH (2013) Some observations in the vortex-turning behaviour of noncircular inclined jets. Exp Fluids 54:1–11

Shi S, Wang J, Ding J et al (2016) Parametric study on light field volumetric particle image velocimetry. Flow Meas Instrum 49:70–88

Soria J (1996) An investigation of the near wake of a circular cylinder using a video-based digital cross-correlation particle image velocimetry technique. Exp Therm Fluid Sci 12:221–233

Soria J, Atkinson C (2008) Towards 3C–3D digital holographic fluid velocity vector field measurement—tomographic digital holographic PIV (Tomo-HPIV). Meas Sci Technol 19:74002

Thomason CM, Thurow BS, Fahringer TW (2014) Calibration of a microlens array for a plenoptic camera. In: 52nd aerospace sciences meeting, p 396

Thurow BS, Fahringer T (2013) Recent development of volumetric PIV with a plenoptic camera

Tsai RY (1986) An efficient and accurate camera calibration technique fro 3D machine vision. In: CVPR'86

Wieneke B (2008) Volume self-calibration for 3D particle image velocimetry. Exp Fluids 45:549–556

Wieneke B (2018) Improvements for volume self-calibration. Meas Sci Technol 29:84002

Chapter 5
Simultaneous 3D Surface Geometry and Pressure Distribution Measurement

Shengxian Shi and Mark Kenneth Quinn

Abstract This chapter describes an aerodynamic measurement technique that is capable of simultaneously characterising 3D model geometry and surface pressure distribution with single light-field camera. This technique combines the traditional intensity-based pressure-sensitive paint (PSP) with light-field imaging and requires one additional step for 3D geometry reconstruction. Working procedures of the single light-field camera 3D surface pressure measurement technique will be firstly introduced, followed by details on how to retrieve pressure ratio image and 3D model geometry from raw light-field images. Results on a Mach 5 wind tunnel test will be lastly discussed to demonstrate its measurement performance.

Keywords Light-field PSP · 3D metrology · Surface pressure · Depth estimation · Metric calibration

Introduction

PSP has shown increasing interest and utilisation by the aerospace testing community since early developments in the 1980s and subsequent industrial testing thereafter (Kavandi et al. 1990; Sellers 1995, 2000). The key advantage of PSP as a measurement technique is that it is capable of providing full-surface coverage of pressure, unlike traditional point measurements gathered from transducers. This ability to measure aerodynamic pressure at every location on a complex model has been utilised to great effect in both academic and industrial experimental campaigns.

The technique utilises the phenomenon of oxygen quenching of luminescent molecules. When these oxygen sensitive molecules (known as luminophores) are illuminated with a suitable wavelength of light (normally ultraviolet) they emit a

S. Shi (✉)
School of Mechanical Engineering, Shanghai Jiao Tong University, Shanghai 200240, China
e-mail: kirinshi@sjtu.edu.cn

M. K. Quinn
Department of Mechanical, Aerospace and Civil Engineering, University of Manchester, Manchester M13 9PL, UK
e-mail: mark.quinn@manchester.ac.uk

S. Shi and T. H. New (eds.), *Development and Application of Light-Field Cameras in Fluid Measurements*, https://doi.org/10.1007/978-3-031-01779-7_5

longer wavelength of light (normally >600 nm). The amount of light emitted is inversely proportional to the concentration of oxygen on the surface and therefore the aerodynamic surface pressure. If such emission is imaged by a suitable camera and processed with sophisticated algorithms, each pixel functions like a pressure tap and the pressure measurement resolution can be as high as the image sensor (Bell et al. 2001; Liu and Sullivan 2005). With suitable pressure sensitive paint and cameras, PSP can achieve accurate measurement for both steady pressure and unsteady pressure fluctuations (Gregory et al. 2008; Peng and Liu 2020). The main sources of uncertainty associated with PSP measurements have traditionally been PSP temperature sensitivity, model displacement and deformation, instability of input light and PSP photo bleaching. The technique has evolved significantly over the past three decades to mitigate temperature sensitivity through advances in paint chemistry (Gouin and Gouterman 2000; Puklin et al. 2000), and through the use of modern LEDs and drivers which provide a very stable input signal to the PSP coating. However, issues of model deformation and deflection have been harder to solve.

Researchers have measured surface shape and pressure using photogrammetry (Crafton et al. 2005; Gramola et al. 2020), digital image correlation techniques (Lynch et al. 2018; Ogg et al. 2018; Quinn and Fisher 2020; Imai et al. 2021), however, each of these experimental campaigns required the use of multiple cameras calibrated in space adding to the complexity of the setup. Especially, mounting and calibration of many cameras for large wind tunnel facilities could be a very troublesome work and will significantly reduce the test efficiency. (Li et al. 2021) demonstrated a single camera system to measure surface shape and pressure utilising a phase-shifting input light; however, the results showed significant distortions in the PSP results due to the spatial profile of the incoming light.

Pressure Estimation from Central Perspective Images

The light-field PSP technique mainly consists two steps, capturing wind-on, wind-off pressure image (Fig. 5.1a), and 3D model geometry by light-field camera (Fig. 5.1b). The first step is very similar to traditional PSP technique except that conventional camera is replaced by a light-field camera. The second step is accomplished with the help of structured light illumination, which helps to increase surface texture.

According to the intensity-based approach, pressure signals can be derived from the ratio of wind-on and wind-off images (Liu and Sullivan 2005). Wind-on image refers to the image that is taken for the UV excited model when wind tunnel is running at certain speed, while wind-off image is normally captured right after the wind tunnel is turned off. The principle behind such processes can be modelled as the following Stern–Volmer equation.

$$\frac{I_{\text{ref}}}{I} = A(T) + B(T)\frac{P}{P_{\text{ref}}} \tag{5.1}$$

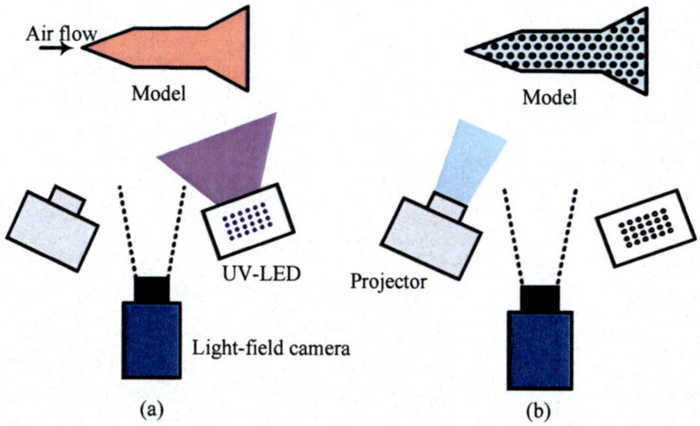

Fig. 5.1 Schematic of the light-field PSP technique, **a** wind-on and wind-off pressure image acquisition with light-field camera. **b** 3D model geometry capturing with light-field camera and structured light illumination

in which I_{ref} is the pixel intensity of images captured at wind-off condition; I is the pixel intensity of images captured at wind-on condition; $A(t)$ and $B(t)$ are paint calibration coefficients which are normally dependent on local temperature; P is the model surface pressure distribution at wind-on condition; P_{ref} is the model surface pressure distribution at wind-off condition (standard atmospheric pressure).

Unlike traditional intensity-based PSP technique where image ratio can be directly taken for wind-on and wind-off images, the raw wind-on and wind-off light-field images need to be pre-processed as they capture both spatial and intensity information of the pressure signals (Fig. 5.2). Such pre-process includes MLA calibration and central perspective calculation. Similar as what detailed in Chapters 2 and 3, MLA calibration is to determine centre coordinate for each lenslet, which is achieved by adjusting the main lens aperture to its minimum and taking light-field image for a white plate. Figure 5.3a shows one of the MLA calibration images, where each white point is the centre of a lenslet. From this image, centre coordinate of each lenslet (red asterisk in Fig. 5.3b) can be calculated by the 3×3 Gaussian fit method with sub-pixel accuracy.

The second step of pre-process is to calculate new perspective images, particularly important for image ratio calculation is to derive central perspective image from the raw wind-on and wind-off light-field images. To simplify the illustration, we scale down the resolution of MLA to 5×5, and pixel resolution behind each lenslet is correspondingly reduced to 5×5 (Fig. 5.4a). Based on light-field imaging principle, such configuration means 25 new perspective images can be calculated from one raw light-field image. Notice that light-field cameras involved in this book have much higher MLA and image sensor resolution. With the known lenslet centre coordinate, pixels that covered by each lenslet can then be precisely identified, and can be rearranged for generating new perspective images. For instance, if all "blue" pixels in Fig. 5.4a (the fifth pixel beneath each lenslet) are selected and arrange into a 5×5

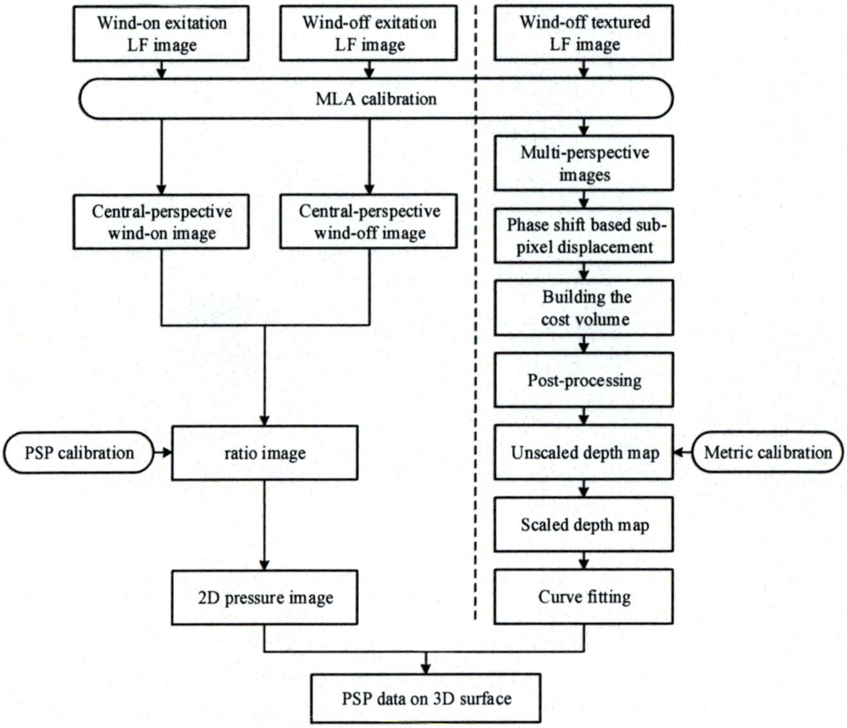

Fig. 5.2 Data processing procedures of the light-field PSP technique

matrix (Fig. 5.4c), a new perspective that views the object from upper-most angle can be generated. Likewise, if all "red" pixels in Fig. 5.4a (the third pixel beneath each lenslet) are selected and combined together, a central perspective image can be generated. For intensity ratio based pressure estimation, only central perspective images are used.

3D Model Geometry Reconstruction from Single Light-field Image

As shown in the right hand side of Fig. 5.2, an additional wind-off light-field image is taken with the aid of "pattern illumination", which is produced by projecting a dense dotted-array image onto the model surface (Fig. 5.1b). The raw wind-on and wind-off light-field images maybe used for model 3D geometry measurement, but they likely result in unreliable data as PSP paint tends to smooth the model surface and very few texture could be used for depth estimation algorithms. By artificially creating a dotted-array pattern on model surface, the wind-off textured light-field image can then be processed with traditional light-field depth algorithm (Jeon et al. 2015). To do that, different perspective images (e.g. 5 × 5 new perspectives in Fig. 5.4) are firstly extracted and built into an image matrix. As illustrated in Fig. 5.5a, new perspective images are arranged clockwise around the central perspective. Depth information between different perspectives can then be calculated with sub-pixel perspective-shift method. For instance, the upper right perspective image (e.g. the red circle image highlighted in blue square) is shifted with different sub-pixel displacements (e.g. 64 labels, Fig. 5.5b) via image interpolation. By comparing the similarity between shifted images with the central perspective one (e.g. the green circle image highlighted in blue square), a series of cost volume will be calculated (Fig. 5.5c). Lastly, the least cost value reveals the parallax between different perspective images, and its corresponding label is determined as depth information. Repeating such calculation for all perspectives, a depth image can be obtained which quantifies the relative position of a pixel to the camera focal plane. To convert such depth image into 3D point cloud, a metric calibration is necessary.

Metric calibration is a process that establishes the mapping between light-field camera coordinate (depth image) with real world coordinate. A typical setup of a metric calibration system in shown in Fig. 5.6a, it consists of a dotted-array or checkerboard (Fig. 5.6b), a high-precision linear translation system and a light-field camera. To perform metric calibration, a calibration board with size similar as measurement object is incrementally translated to and fro along the light-field camera optical axes. For each position, light-field images are taken (Fig. 5.6c) and processed by the depth estimation method. With the calculated depth image and the known spatial coordinates of dotted-array/checkerboard, the mapping function can be established via spline fit. Finally, depth image of a model can be converted into metric 3D point cloud (3D geometry) with such mapping function.

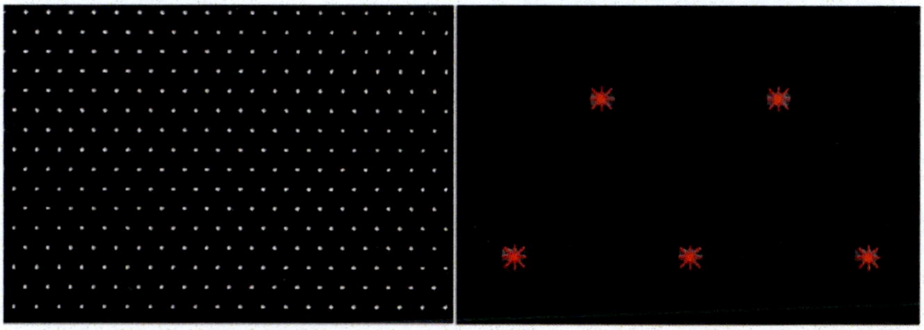

(a) (b)

Fig. 5.3 **a** Light-field image of white plate with main lens aperture set to minimum, **b** lenslet centre determined with Gaussian fit method

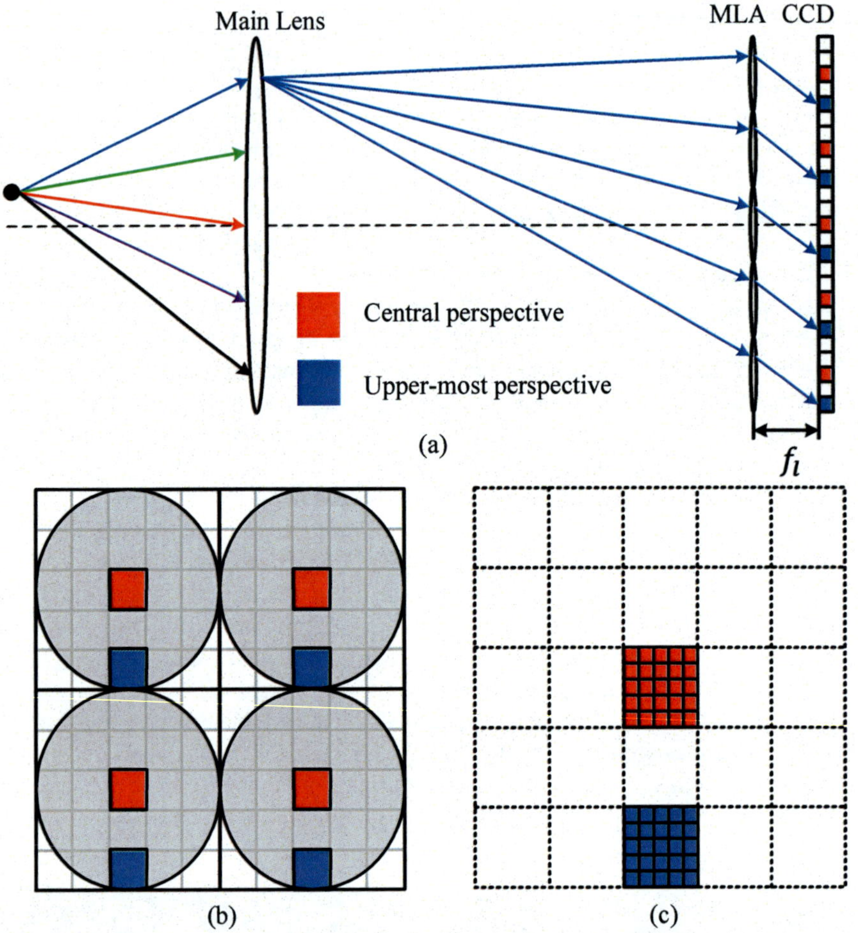

Fig. 5.4 **a** Principle of new perspective image generation, **b** example of lenslet and pixels behind it, **c** new perspective images generation for central and upper-most viewpoints

Application of Light-field PSP to Mach 5 Wind Tunnel Experiment

The light-field PSP technique has been demonstrated in the intermittent blow-down high supersonic tunnel (HSST) at the Department of Mechanical, Aerospace and Civil Engineering at the University of Manchester. The HSST, shown in Fig. 5.7a, is of a free jet design which expands into a quiescent plenum chamber held nominally at vacuum. The HSST can provide 7.5 s of stable run time due to the large high-pressure storage tanks. Experiments were performed at Mach 5 with a maximum total pressure of 8.1 bar, while the total temperature was maintained at 300 K via a resistive heater. Previous tests around this total temperature have shown that, following nozzle

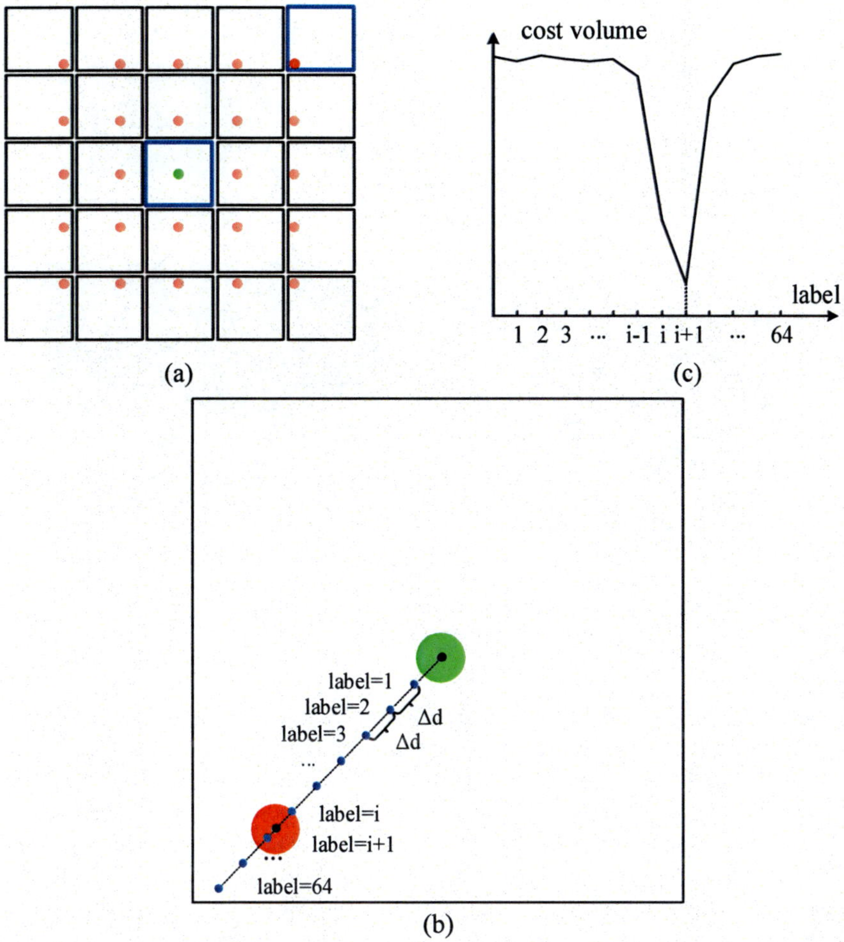

Fig. 5.5 Principle of the depth estimation method, **a** 5 × 5 perspective images are arranged clockwise around the central perspective, **b** shift perspective image with sub-pixel displacement and compare it with central image, **c** calculate cost volume (similarity) for each label and determine depth value

expansion and subsequent shock recompression, the static temperature change on the surface is negligible between wind-on and wind-off conditions (Quinn and Fisher 2020). There are large quartz windows at two sides of the test section to facilitate Schlieren and PSP measurements. To provide a qualitative comparison for light-field PSP data, Schlieren experiments were performed with a *Z*-style Töepler system, where illumination was provided by a Newport-Oriel 1 kW Xenon Arc Lamp and Schlieren images were captured by a Nikon D5000 SLR camera.

A flared cone model was painted with ISSI UniFIB PSP paint and illuminated by an in-house UV LED arrays (Quinn et al. 2017). A calibration curve for the PSP was

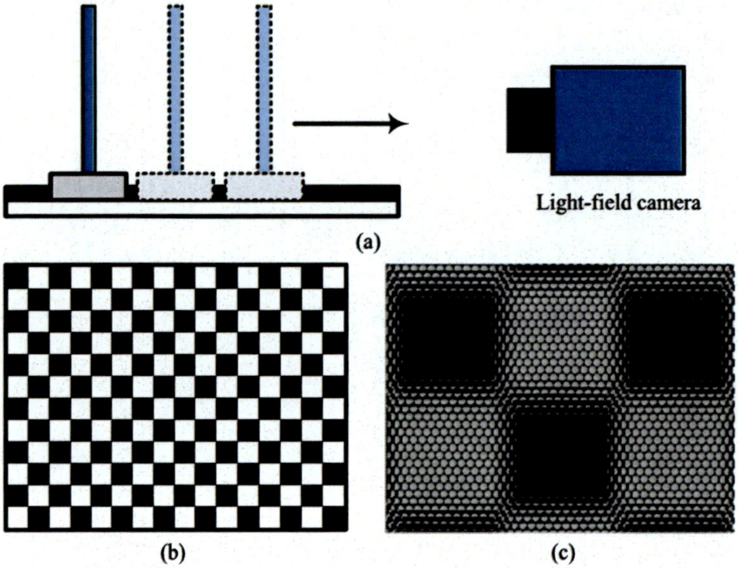

Fig. 5.6 **a** Sketch of the metric calibration system, **b** the checkerboard for light-field camera calibration, **c** raw light-field image of the checkerboard

generated using an *a*-priori calibration chamber as described by Quinn et al. (2017). Pressure and temperature were varied independently in the calibration chamber with identical lighting and optical components to produce a response function for the PSP, enabling quantitative results. Figure 5.7b shows the geometry used in this experiment, which is the same as the one studied by Erdem (2011) and Erdem et al. (2011).

During the experiments, three sets of light-field images were captured by the in-house 29 M light-field camera for the flared cone model, which are wind-on light-field images, wind-off light-field images and wind-off light-field images with dotted-array pattern illumination (wind-off textured light-field image). For the wind-on and wind-off light-field images, a 610 nm long-pass filter was mounted in front of the main lens to block the excitation light. Figure 5.8 shows an example for these three image sets, where a clear increase in pixel intensity can be seen in Fig. 5.8a, b due to the surface pressure variation, and dotted-array pattern can be seen from Fig. 5.8c. Process such images with procedures detailed in Fig. 5.2, central perspective images were extracted from raw wind-on and wind-off light-field images, similar as traditional PSP, these central perspective wind-on and wind-off images were then taken for the ratio calculation and eventually estimate pressure distribution with the aid of PSP calibration data.

Figure 5.9 shows the normalised model surface pressure distribution, Schlieren image, and pressure profile along the model centreline. A distinct asymmetry of pressure distribution can be observed at the flared region (Fig. 5.9a). This is most likely resulted by model rotation at Mach 5 flow, which results in a minor negative incidence angle. Such rotation effect was magnified at the model trailing edge. The

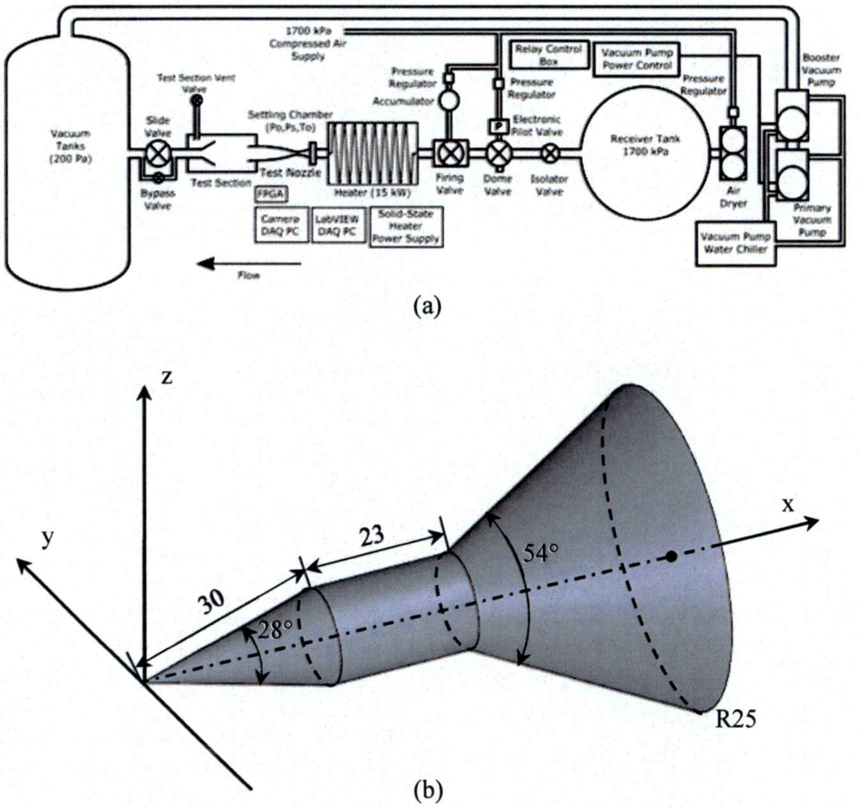

Fig. 5.7 **a** Sketch of the supersonic wind tunnel experiment system. **b** 3D drawings of flared cone model (in mm)

asymmetry error could be corrected if image correction between wind-on and wind-off were taken. Apart from this minor defect, Fig. 5.9c clearly shows that the model tip compressed incoming flow from static pressure to a value of $P/P_\infty \approx 3.4$, which matches well with the theoretical value for a Mach 5 flow around a 14° cone. Flow then separated at $X/L \approx 0.36$ due to the change to constant radius, as reflected by both pressure profile and Schlieren results (Fig. 5.9b, c). The separated flow subsequently formed shear layer impinged on the flared region at $X/L \approx 0.70$, and eventually produced a strong shockwave. Such observations agree well with early studies (Erdem et al. 2011). Furthermore, pressure distribution along centreline matches well with CFD studies made by Erdem (2011), with only difference that the measured highest pressure is larger than predicted by CFD. This could be induced by minor temperature changes during experiments, and maybe corrected with two component PSP.

On the other hand, new perspective images were calculated from raw wind-off textured light-field image. The angular resolution of in-house light-field camera is

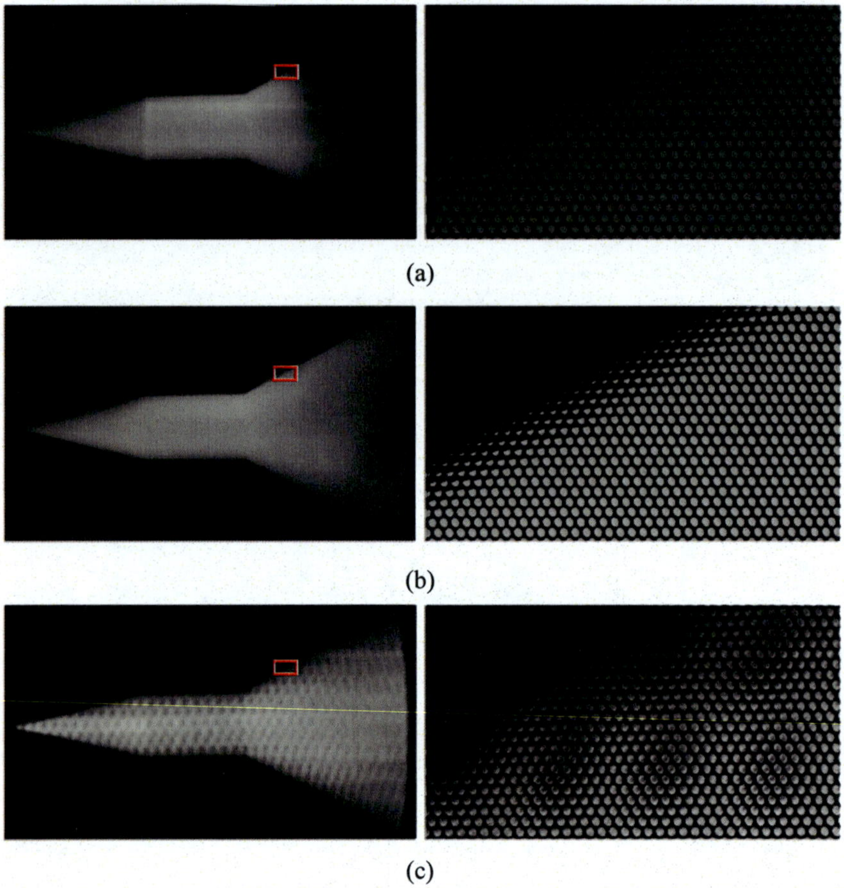

Fig. 5.8 **a** Raw light-field image of the flared cone model at wind-on condition, **b** raw light-field image of the flared cone model at wind-off condition, **c** raw light-field image of the flared cone model under the illumination of dotted-array pattern

14×14, thereby 14×14 new perspective images can be generated from one raw wind-off textured light-field image. To reduce the computational load, only the central 5×5 perspective images were involved in depth image calculation. Figure 5.10 shows the calculated 3D geometry after calibrating the depth image. To evaluate the measurement accuracy, 3D geometry data were extracted for three cross sections at $x = 89.5$ mm, $x = 47.0$ mm, and $x = 24.5$ mm (as indicated by dashed lines in Fig. 5.10), these discrete data along with the fitted curves were plotted in Fig. 5.11a. It demonstrates that the measured 3D data generally match with the model surface curvature. For a more direct comparison, the model design drawings are served as reference data (machining error of CNC is around $\pm 20\,\mu$m), absolute 3D measurement errors are calculated along $y = 30.0$ mm. As shown in Fig. 5.11b, measurement errors in regions $x = 0$–30 mm are relatively high, especially near the model tip. In

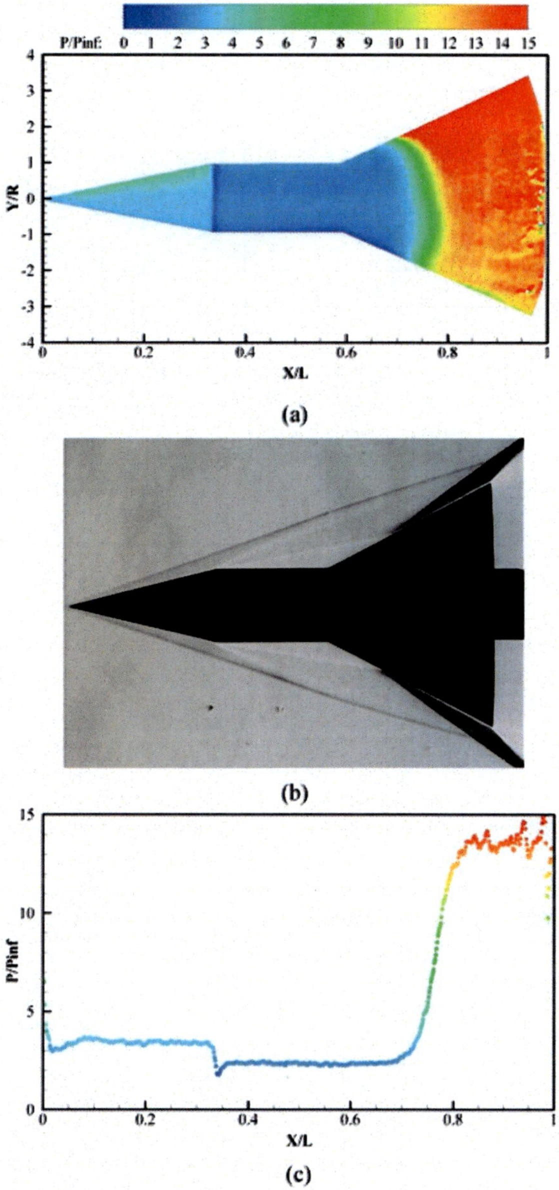

Fig. 5.9 **a** Normalised model surface pressure distribution calculated from central perspective wind-on and wind-off images, **b** Schlieren image of the test model at same test conditions, **c** pressure profile along the model centreline ($Y/R = 0$)

these areas, dotted-pattern cannot provide enough texture features for accurate depth calculation, due to very small radius. For regions $x = 30{-}90$ mm, where radius are relatively large and there are enough dotted-pattern scattered on the model surface, the 3D measurement errors are less than 1 mm.

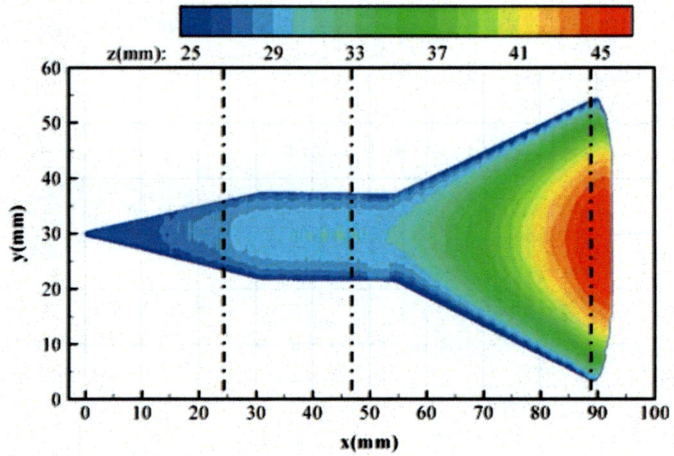

Fig. 5.10 3D geometry of the flared cone model

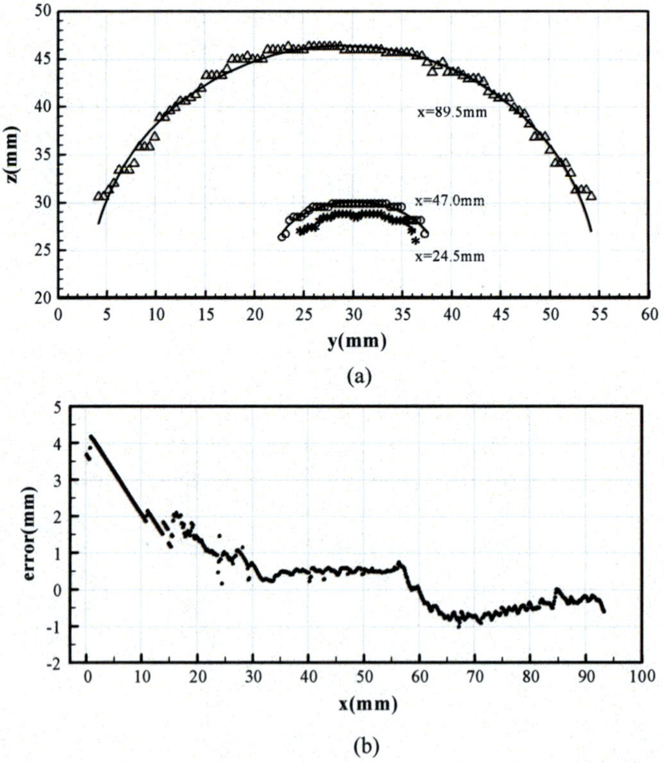

Fig. 5.11 **a** 3D geometry data at $x = 89.5$ mm, $x = 47.0$ mm and $x = 24.5$ mm. **b** 3D measurement error along $y = 30$ mm

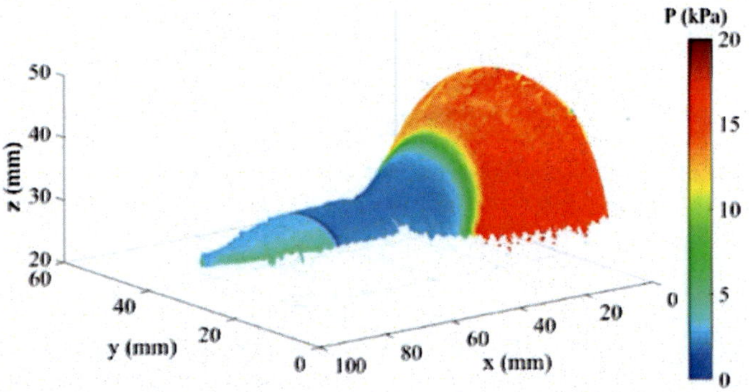

Fig. 5.12 3D surface pressure distribution of the flared cone model

At this point, the 3D model data along with pressure distribution are both acquired, and are at the same resolution as central perspective image. Therefore, it is just a matter of pixel-to-pixel mapping to merge the 3D geometry data with surface pressure data, and producing the finally 3D surface pressure distribution (Fig. 5.12).

References

Bell JH, Schairer ET, Hand LA, Mehta RD (2001) Surface pressure measurements using luminescent coatings. Annu Rev Fluid Mech 33:155–206

Crafton J, Fonov S, Jones E et al (2005) Simultaneous measurements of pressure and deformation on a UCAV in the SARL. In: 43rd AIAA aerospace sciences meeting and exhibit, p 1028

Erdem E (2011) Active flow control studies at Mach 5: measurement and computation

Erdem E, Yang L, Kontis K (2011) Drag reduction studies by steady energy deposition at Mach 5. In: 49th AIAA aerospace sciences meeting including the new horizons forum and aerospace exposition, p 1027

Gouin S, Gouterman M (2000) Ideality of pressure-sensitive paint. II. Effect of annealing on the temperature dependence of the luminescence. J Appl Polym Sci 77:2805–2814

Gramola M, Bruce PJ, Santer MJ (2020) Response of a 3D flexible panel to shock impingement with control of cavity pressure. In: AIAA scitech 2020 forum, p 314

Gregory JW, Asai K, Kameda M et al (2008) A review of pressure-sensitive paint for high-speed and unsteady aerodynamics. Proc Inst Mech Eng Part G J Aerosp Eng 222:249–290

Imai M, Nakakita K, Nakajima T, Kameda M (2021) Unsteady surface pressure measurement of transonic flutter using a pressure sensitive paint with random dot pattern. In: AIAA scitech 2021 Forum, p 127

Jeon H-G, Park J, Choe G et al (2015) Accurate depth map estimation from a lenslet light field camera. In: Proceedings of the IEEE conference on computer vision and pattern recognition, pp 1547–1555

Kavandi J, Callis J, Gouterman M et al (1990) Luminescent barometry in wind tunnels. Rev Sci Instrum 61:3340–3347

Li Y, Dong Z, Liang L et al (2021) Simultaneous 3D surface profile and pressure measurement using phase-shift profilometry and pressure-sensitive paint. Rev Sci Instrum 92:35107

Liu T, Sullivan JP (2005) Pressure and temperature sensitive paints. Springer, Berlin

Lynch KP, Jones E, Wagner JL (2018) Simultaneous PSP and surface deformation measurements for fluid-structure interactions in a shock tube. In: 2018 fluid dynamics conference, p 3870
Ogg DR, Rice BE, Peltier SJ et al (2018) Simultaneous stereo digital image correlation and pressure-sensitive paint measurements of a compliant panel in a Mach 2 wind tunnel. In: 2018 fluid dynamics conference, p 3869
Peng D, Liu Y (2020) Fast pressure-sensitive paint for understanding complex flows: from regular to harsh environments. Exp Fluids 61:1–22
Puklin E, Carlson B, Gouin S et al (2000) Ideality of pressure-sensitive paint. I. Platinum tetra (pentafluorophenyl) porphine in fluoroacrylic polymer. J Appl Polym Sci 77:2795–2804
Quinn MK, Fisher TB (2020) Simultaneous measurement of surface shape and pressure using structured illumination. Meas Sci Technol 32:24011
Quinn MK, Spinosa E, Roberts DA (2017) Miniaturisation of pressure-sensitive paint measurement systems using low-cost, miniaturised machine vision cameras. Sensors 17:1708
Sellers ME (1995) A comparison of an AEDC and a Russian developed pressure sensitive paint in the AEDC propulsion wind tunnel 16T
Sellers ME (2000) Application of pressure sensitive paint for determining aerodynamic loads on a scale model of the F-16C. In: 21st aerodynamic measurement technology and ground testing conference, p 2528

Chapter 6
Light-field PIV Implementation and Case Studies

T. H. New, Shengxian Shi, J. Soria, and B. Ganapathisubramani

Abstract This chapter will provide details on how LF-PIV had been applied for both fundamental and more applied aerodynamics and fluid dynamics studies, covering significantly different flow conditions and test mediums. These case studies will shed light upon how LF-PIV may be implemented in different experimental setups and how it compares against conventional 2D- or Tomo-PIV approaches. Another purpose here is to highlight the advantages, disadvantages and how LF-PIV implementations could be beneficial in more challenging experimental setups or flow scenarios.

Keywords Light-field PIV · Turbulent boundary layer · Impinging jet · Linear cascade · Supersonic jet

Introduction

The implementation and use of LF-PIV with satisfactory accuracy levels and at reasonable computational costs have only been established fairly recently. This

T. H. New (✉)
School of Mechanical and Aerospace Engineering, Nanyang Technological University, Singapore 639798, Singapore
e-mail: DTHNEW@ntu.edu.sg

S. Shi
School of Mechanical Engineering, Shanghai Jiao Tong University, Shanghai 200240, China
e-mail: kirinshi@sjtu.edu.cn

J. Soria
Laboratory for Turbulence Research in Aerospace and Combustion, Department of Mechanical and Aerospace Engineering, Monash University, Melbourne, VIC 3800, Australia
e-mail: julio.soria@monash.edu

B. Ganapathisubramani
Aerodynamics and Flight Mechanics Research Group, University of Southampton, Southampton SO17 1BJ, UK
e-mail: G.Bharath@soton.ac.uk

S. Shi and T. H. New (eds.), *Development and Application of Light-Field Cameras in Fluid Measurements*, https://doi.org/10.1007/978-3-031-01779-7_6

is in no small part due to rapid advancements in camera sensor quality, resolution, data readout speed, MLA, Graphics Processing Unit (GPU)-aided post-processing through the use of Nvidia CUDA cores and Scalable Link Interface (SLI), among others. However, powerful hardware alone is insufficient and advanced post-processing algorithm is necessary to accelerate the process and reduce the computational time to an acceptable level. This is quite similar to the situation when 2D and Tomo-PIV techniques were first introduced many years ago, when post-processing in the former sometimes require acceleration from dedicated external hardware connected to the workstation. For the LF-PIV implementations used by the authors in this book, post-processing acceleration can be achieved much more readily without resorting to the use of specially-designed hardware but instead, by tapping into the relatively low-cost but highly powerful, off-the-shelf CUDA-core based GPU cards. A combination of DRT-MART particle reconstruction approach and multiple GPUs implemented in parallel operations, as described and discussed in earlier chapters, allow significantly different flow scenarios to be investigated by LF-PIV recently. This chapter aims to provide some brief details on selected investigated flow scenarios in terms of the implementation and the results obtained, so as to provide readers a sense of what is possible with LF-PIV and the scientific insights that are achievable. These are by no means an extensive coverage and readers are encouraged to refer to the relevant publications for much more details.

Adverse Pressure-Gradient Turbulent Boundary Layers

A flow scenario for which LF-PIV had been implemented for will be a turbulent boundary layer under self-similar adverse pressure gradient (APG-TBL) conditions, as reported by Zhao et al. (2019). This series of investigations was conducted in the water tunnel at the Laboratory for Turbulence Research in Aerospace and Combustion (LTRAC), Monash University and the purpose was to compare and validate LF-PIV measurements with conventional 2D-PIV measurements of the present flow scenario. As Fig. 6.1 shows, an additional contraction section and a flexible polycarbonate roof within the test-section was implemented to produce the APG, where the APG could be adjusted by varying the roof height, and hence the flow cross-sectional dimensions, continuously along the test-section. To achieve the desired APG, the boundary layer thickness above a glass-based false floor positioned within the desired test location was firstly measured at several locations along the streamwise direction using 2D-PIV approach. The captured images were then processed in real-time using an in-house image acquisition and PIV analysis code, with the results used to adjust the flexible roof iteratively (Atkinson et al. 2015) to obtain the desired APG. Tests showed that a self-similar APG-TBL was formed at the fourth section of the water tunnel under moderate APG conditions (Atkinson et al. 2015, 2016) and that location was used for the rest of the present study.

The non-dimensionalised APG is defined as

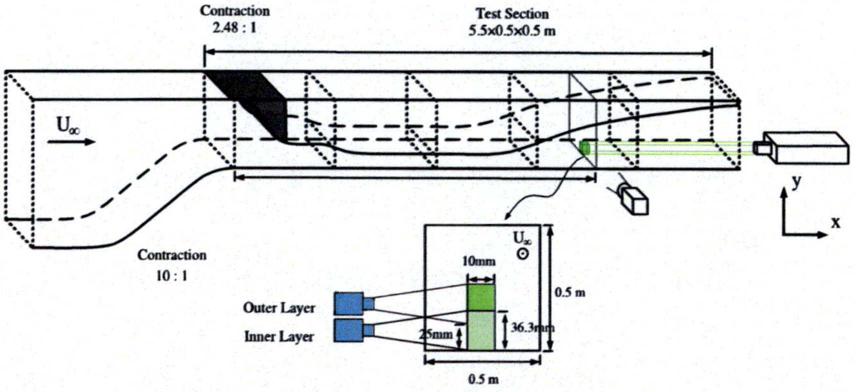

Fig. 6.1 Schematics of the water tunnel and LF-PIV implementation

$$\beta = \left(\frac{\delta_1}{\tau_w}\right)\frac{\mathrm{d}P}{\mathrm{d}x} \tag{6.1}$$

where $\mathrm{d}P/\mathrm{d}x$ is the streamwise pressure gradient, δ_1 is the boundary layer displacement thickness, τ_w is the local mean wall shear stress, and can be worked out to be approximately 2.0 ± 0.15. The averaged external flow velocity within the self-similar region was estimated to be about 440–470 mm/s, while the Reynolds number at the entrance of this region is about $\mathrm{Re}_{\delta 1} = 5400$.

Before any LF-PIV measurements were conducted however, 2D-PIV measurements were carried out first to provide the necessary datasets to validate the implementation of LF-PIV for this particular flow scenario. Characteristics and self-similarities of the APG-TBL were captured and studied further at several streamwise locations from $x = 3.68$–4.18 m at 100 mm intervals. Since only velocity profiles were needed, a high-speed sCMOS camera binned at a resolution of 96×2560 pixels was used to capture the 2D particle images. Note that the 96-pixel value corresponds to the horizontal (i.e. streamwise) direction, while the 2560-pixel is the full camera vertical resolution, since the velocity profiles in the vertical (i.e. normal to the wall) direction were to be measured. Illumination was provided by a 1 mm thick laser sheet formed by a 22mW, 532 nm wavelength continuous-wave laser. An exposure time of 0.3 ms was used to capture the 11 μm hollow glass sphere particles dispersed within the water tunnel. The camera acquisition frequency was 500 Hz and the captured particle images were post-processed using 2D multi-grid cross-correlations with initial and final interrogation windows of 96×64 pixels and 64×24 pixels respectively. The settings gave rise to a vector resolution of 0.012δ in the direction normal to the wall. Based on the wall-normal velocity profile, the wall shear stress and hence, non-dimensionalised streamwise pressure gradient, could subsequently be calculated.

As for LF-PIV, the experimental arrangement can be understood in Fig. 6.1, where the light-field camera was positioned such that its optical axis is normal to

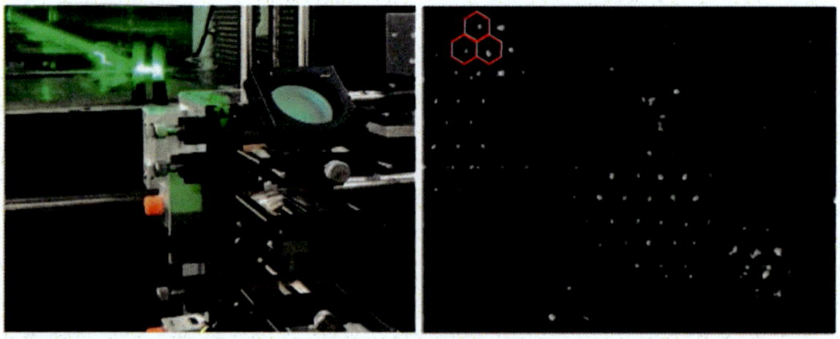

(a) (b)

Fig. 6.2 **a** Photo of the LF-PIV system implemented next to the water tunnel and a **b** partially magnified particle image. The MLA arrangement can be seen in the red hexagonal frame in (**b**), where each cell is a single micro-lens that collects a sub-image

the test-section side wall (see Fig. 6.2a). There were two measurement volumes—one for the near-wall region (i.e. inner layer) and another for the external part of the boundary layer (i.e. outer layer), with the light-field camera location adjusted for these two different measurement volumes. Similar to the 2D-PIV measurements, 11 μm hollow glass spheres were used as seeding particles and a 90 mJ/pulse, 532 nm wavelength, double-pulsed Nd:YAG laser was used as the illumination source. As shown in the figure, before the laser entered the water tunnel horizontally from the end wall, the former was first formed into a laser volume through the use of two cylindrical lenses. The same in-house light-field camera modified from an Imperx B6640 PIV camera with a 520 $\times$ 360 MLA was used here with a Micro-Nikkor 200 mm lens. Magnification factor used was 1 for optimal accuracy and the particle density was about 0.5PPM. The time-interval between particle image-pairs was maintained at 0.6 ms throughout the LF-PIV experiments and 600 image-pairs were captured for each measurement volume. To post-process the light-field images (an example is shown in Fig. 6.2b), DRT-MART approach with pixel-to-voxel ratios of 3, 3 and 10 in the x, y and z directions respectively were utilised. This gave a reconstruction domain size of 733 $\times$ 2200 $\times$ 182 voxels and spatial resolution of 0.0165 $\times$ 0.0165 $\times$ 0.055 mm^3/voxel. Reconstructed 3D particle images were then subjected to 3D multi-grid cross-correlations with initial and final interrogation volumes of 256 $\times$ 256 $\times$ 64 pixel and 128 $\times$ 128 $\times$ 32 pixel respectively. After appropriate vector validations, the 3D velocity fields would have a resolution of 0.558 $\times$ 0.637 $\times$ 0.526 mm^3/vector, for a total of 23 465 (19 $\times$ 65 $\times$ 19) velocity vectors per velocity field.

To show how the individual result of the two different measurement volumes look like, Fig. 6.3a and b show instantaneous 3D velocity fields worked out for the two measurement volumes. Note that the water tunnel wall is oriented to the right of the results. Since they overlapped, 600 instantaneous velocity fields were averaged for each measurement volume, before been combined into a single mean 3D velocity field shown in Fig. 6.3c. While the mean 3D velocity field looked reasonable and logical, a more systematic assessment of all the collected instantaneous velocity fields were still carried out. This was done by examining the flow divergence for

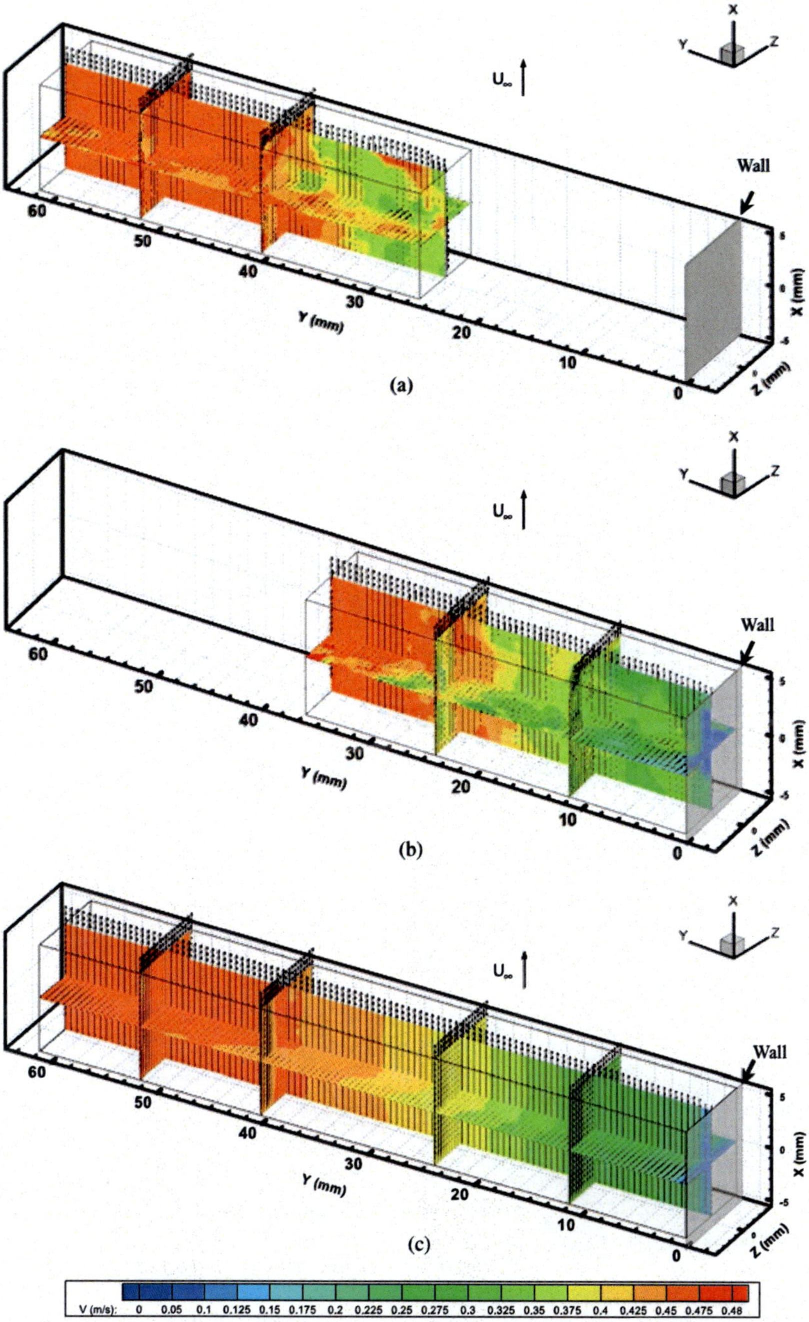

Fig. 6.3 Instantaneous 3D velocity fields obtained for the **a** outer layer and **b** inner layer, as well as the **c** combined mean 3D velocity field. Wall is oriented to the right vertical side of the plots

incompressible flows based on the continuity equation, expressed as

$$\frac{\partial u}{\partial x} + \frac{\partial v}{\partial y} + \frac{\partial w}{\partial z} = 0 \quad (6.2)$$

As such, the spanwise velocity gradients were plotted against the sum of the streamwise and wall-normal velocity gradients and presented as joint probability distribution functions (JPDF) in Fig. 6.4. For perfectly accurate measurements, the above continuity equation must be satisfied and the distributions will be represented by the red dashed lines in the plots. However, for real-world flow measurements, the distributions will not lie perfectly along those lines but scattered about an area.

It should be mentioned that the nature of the data scatter observed in the JPDFs will shed light on the measurement accuracy with respect to the x, y and z directions. With reference to Fig. 6.4a and c, the more elongated distribution of data scatter along the $\partial w/\partial z$ axis as compared to the $\partial u/\partial x + \partial v/\partial y$ axis indicates that the resolution in the

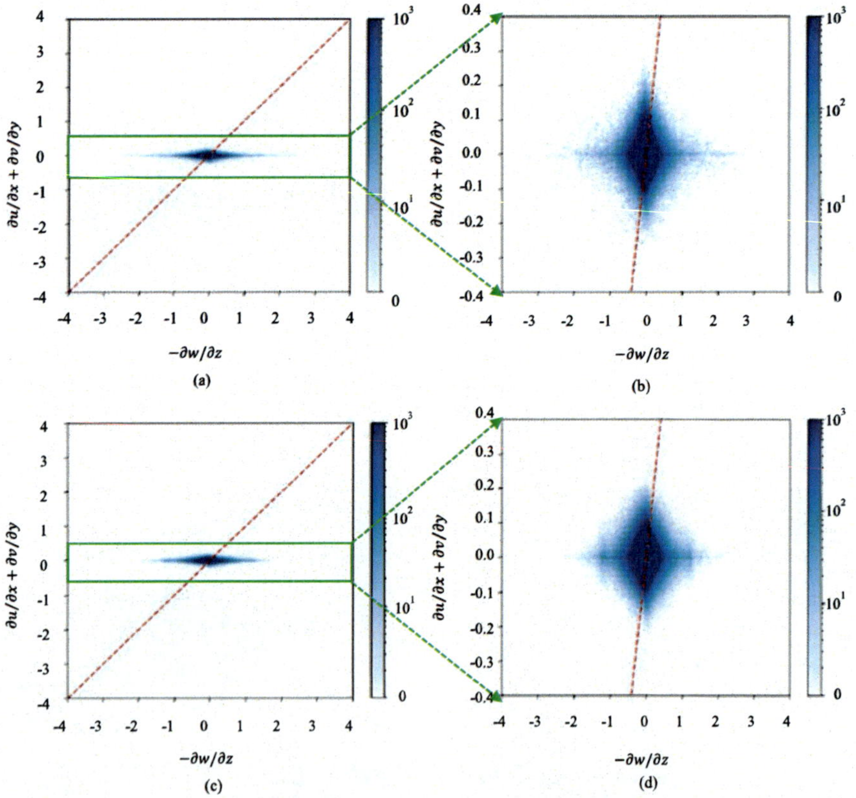

Fig. 6.4 Divergence test result plots for the **a** outer layer, where **b** shows a close-up view of the green box in (**a**). Similar divergence test result plots for the **c** inner layer and the corresponding **d** close-up view of the green box in (**c**) are also included here for comparison

z direction is not as high as those in the x and y directions, since the fluctuations and hence uncertainties are significant higher for $\partial w/\partial z$. However, this is unsurprising since the z direction is along the light-field camera optical axis and represents the depth direction for the light-field camera. As we had seen in earlier chapters, the resolution in the depth direction for LF-PIV approach is known to be worse than those in the other two orthogonal directions (Deem et al. 2016; Shi et al. 2016). Nevertheless, the data scatter regions remain centred and practically symmetrical about the origin locations. To take a closer look, Fig. 6.4b and d show the close-up views of the data scatter, while rescaling the axes such that the $\partial w/\partial z$ axis are now scaled 10 times that of the $\partial u/\partial x+\partial v/\partial y$ axis. While this makes the data scatter more evenly distributed about the origin location, it also better reveals that the uncertainties in the measured velocity gradients in the z direction are substantially larger (i.e. ~ 10 times) than those in the x and y directions. Considering that pixel-to-voxel ratios in the x, y and z directions are 3, 3 and 10 respectively, the use of a larger value in the z direction does not significantly reduce the measurement uncertainty in that direction.

Next, comparisons of the mean statistical profiles obtained by 2D-PIV and LF-PIV approaches in both the outer and inner layers are presented in Fig. 6.5. For the streamwise velocity profile shown in Fig. 6.5a, there are partial agreements between the profiles captured by the two different PIV approaches. Specifically, good agreements can only be found for $y/\delta_1 \approx 1$ and above, with the 2D-PIV and LF-PIV profiles increasingly deviating from each other below that value. There are two primary reasons that can be attributed to this discrepancy that is occuring close to the wall. Firstly, it is known that particle distribution is generally less dense within boundary layers due to the lower speeds encountered. With a lower particle density, the reconstruction quality and subsequent particle identification will subsequently be sub-optimal for LF-PIV. Secondly, note that volumteric calibration (Shi et al. 2019) was not carried out here and hence, no lens imperfections and misalignments between the various optical components were taken into consideration and accounted for. The impact of limited measurement accuracy below $y/\delta_1 \approx 1$ can also be seen in the streamwise and wall-normal components of the Reynold shear stress, where similar limitations in the agreements between 2D-PIV and LF-PIV can be observed. Neverthless, they remain generally acceptable for the streamwise and wall-normal components as shown in Fig. 6.5b and c, where the agreements tend to go below $y/\delta_1 \approx 1$ to about $y/\delta_1 \approx 0.3$ for the inner layer results. Interestingly, Reynolds shear stress results depicted in Fig. 6.5d show abrupt decreases in the magnitudes at locations lower than $y/\delta_1 \approx 0.8$. This could be due to the limited resolution of LF-PIV effecting a spatial filtering like outcome on the small-scale eddies very close to the wall. For the sake of completeness, results for the spanwise component of Reynolds shear stress by LF-PIV approach are provided in Fig. 6.5e. Since 2D-PIV approach does not provide out-of-plane results, no corresponding results for 2D-PIV are available for comparisons here.

The preceding results show that LF-PIV approach possesses the lowest resolution at the focal plane (i.e. $z = 0$) location. If that is the case, it will be informative if results along and some small distance away from the focal plane were to be compared. Figure 6.6 shows such a comparison for streamwise component of Reynolds shear

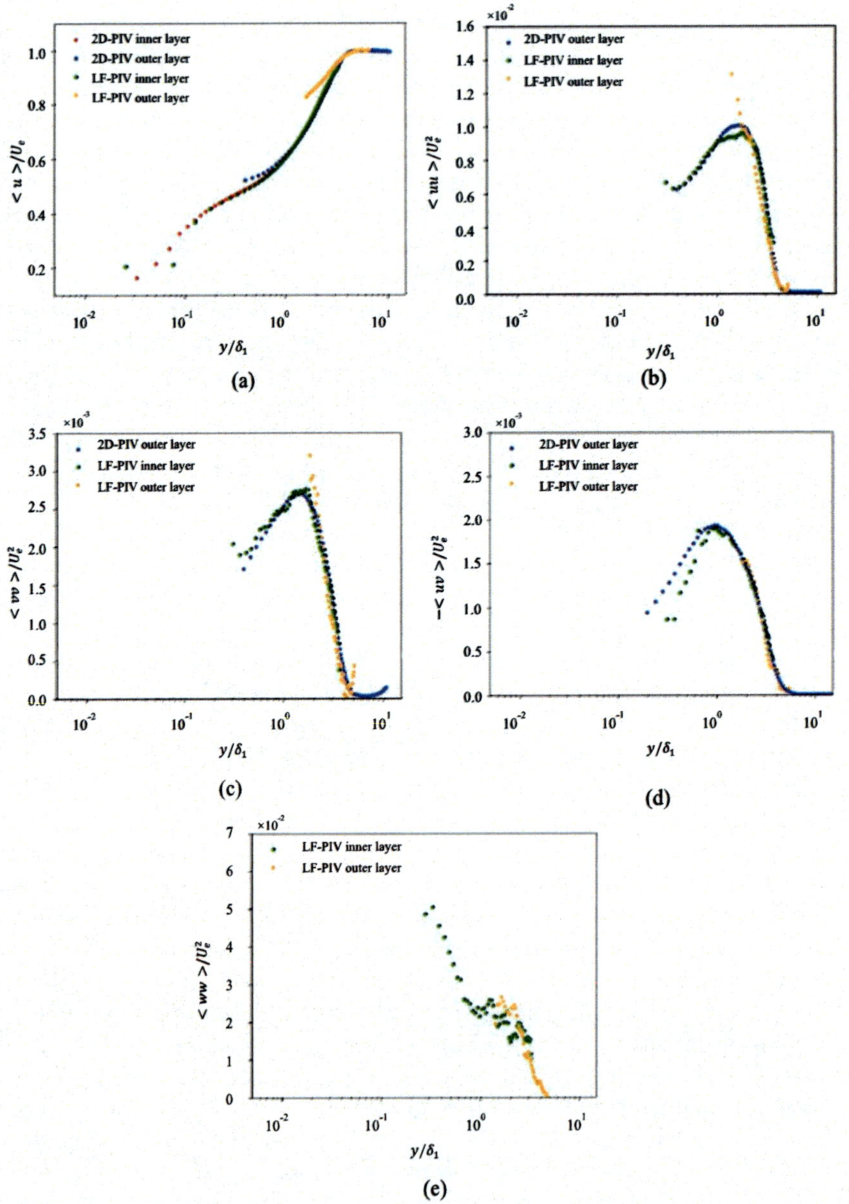

Fig. 6.5 Comparisons of mean profiles obtained by 2D-PIV and LF-PIV for **a** streamwise velocity, **b** Reynolds shear stress streamwise component, **c** Reynolds shear stress wall-normal component and **d** Reynolds shear stress. Reynolds shear stress spanwise component profiles by LF-PIV only are presented in (**e**)

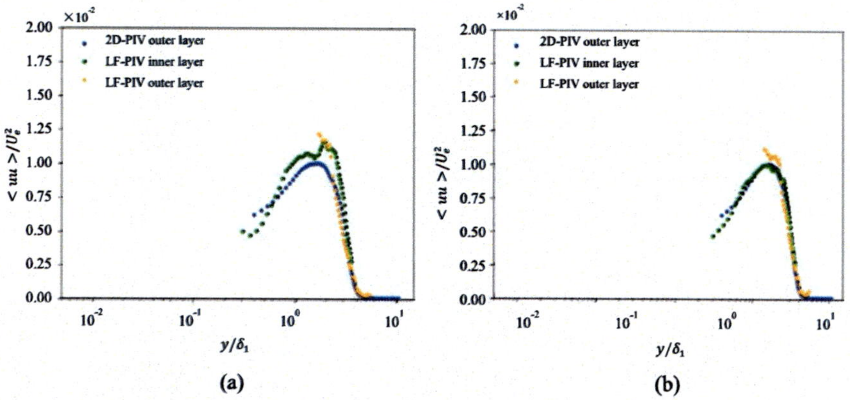

Fig. 6.6 Comparisons of Reynolds shear stress streamwise component along **a** $z = 0$(focal plane) and **b** $z = 1.056$ mm

stress taken at $z = 0$ (focal plane) and 1.056 mm locations to illustrate the outcomes. The comparison reveals that results taken from some small distance away from the focal plane actually lead to better agreements with the 2D-PIV results. This test confirms the notion that while the relatively poorer resolution of LF-PIV along the focal plane may not have a significant impact upon the qualitative aspects of the measured flow, differences (sometimes significant) in the quantitative aspects meant that users of LF-PIV will have to strike a careful compromise between measurement convenience and accuracy of the results.

Impinging Jets

Another area for which LF-PIV technique was used for full 3D flow measurements will be on impinging jets by Zhao et al. (2021). This study is one of the first jet flow related studies making use of single camera LF-PIV, when most earlier studies by the authors involved conventional or time-resolved 2D-PIV and stereoscopic PIV [(Cater and Soria 2002; Ganapathisubramani et al. 2002, 2007, 2008; Lim et al. 2006; New 2009; New and Tsovolos 2009, 2011, 2012; Buxton and Ganapathisubramani 2010; Toh et al. 2010; Shi and New 2013; Edgington-Mitchell et al. 2014a, b; Long and New 2015, 2016, 2019; New and Long 2015; Zang and New 2015, 2017; Berk et al. 2016; Le Clainche et al. 2017), among others].

In this study, a speaker-driven synthetic jet produced from a rectangular orifice was set to impinge upon a flat plate in a relatively confined volume and LF-PIV was used to investigate both the mean and transient aspects of the resulting flow, as shown in Fig. 6.7. The synthetic jet was produced by forcing flows out of a rectangular orifice using a 30 W speaker with the forcing frequency set to $f = 1$ Hz. As such, its centreline velocity at the orifice exit was defined as

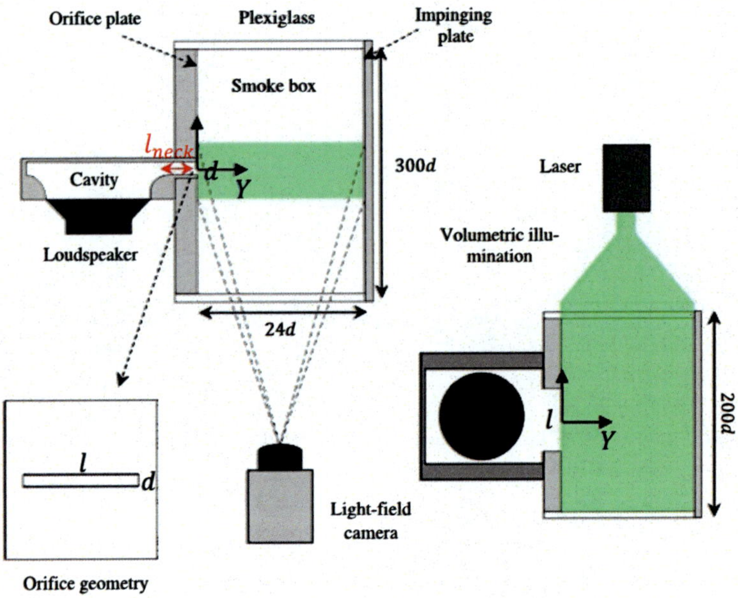

Fig. 6.7 Schematics of the experimental setup for the impinging rectangular synthetic jet

$$u(t) = u_{\max} \sin\left(\frac{2\pi t}{\tau} + \theta\right) \tag{6.3}$$

where $u_{\max}$, t and τ represent the maximum velocity encountered during the blowing cycle, time and period of the cycle, respectively. In addition, the mean blowing velocity, $\overline{u}$, is used as a representative velocity throughout and is defined as

$$\overline{u} = \frac{1}{\tau}\int_0^{\tau/2} u(t)\mathrm{d}t = \frac{1}{\tau}\int_0^{\tau/2} u_{\max}\sin\left(\frac{2\pi t}{\tau}\right)\mathrm{d}t = \frac{u_{\max}}{\pi} \tag{6.4}$$

Since the net centreline flow velocity at the orifice exit will be zero when time-averaged, a momentum flow velocity U_0 will be defined as

$$U_0 = \frac{\overline{u}\pi}{\sqrt{2}} \tag{6.5}$$

Other relevant parameters will include

$$AR = \frac{l}{d} \tag{6.6}$$

$$Sr = \frac{fd}{\overline{u}} = \frac{d}{L_o} \tag{6.7}$$

$$\mathrm{Re} = \frac{\overline{u}d}{\nu} \tag{6.8}$$

$$L_o = \frac{\overline{u}}{f} \tag{6.9}$$

$$f = 1/\tau \tag{6.10}$$

The orifice geometry measured $d = 1$ mm (i.e. width) and $l = 10$ mm (i.e. length), which gave it an aspect ratio of AR = 10. Furthermore, the Reynolds number, Strouhal number and stroke length were worked out to be Re = 550, Sr = 0.0117 and $L_o =$ 0.0855 m respectively. The maximum and momentum flow velocity were determined to be $u_{max} = 26.85$ m/s and $\overline{u} = 8.55$ m/s respectively as well.

With reference to the figure, the speaker was mounted horizontally with a cavity within an enclosure and the synthetic jet was produced horizontally into an enclosed Plexiglas box, where the impingement plate resided. The impingement surface was positioned 24 d away from the jet orifice. 1 μm smoke particles were dispersed within the Plexiglas box with a particle density of about 1 PPM, which were illuminated by a 500 mJ/pulse, 523 nm wavelength, double-pulsed Nd:YAG laser. The illuminated volume was formed through the use of cylindrical convex and concave lenses along the laser beam path. The LF-PIV camera was based on an Imperx B6640 PIV camera with a resolution of 6600 × 4400 pixels incorporating a 520 × 360 array MLA located in front of the imaging sensor. For the double-pulsed measurement experiments, a time interval of 0.1 ms was used so as to satisfy the one-quarter rule for ideal particle displacements between successive particle volumes. For phase-averaged results, a National Instruments NI-PXIe 1062Q data acquisition device was used to trigger the LF-PIV system at 20 different flow phases, with 200 particle image-pairs captured per phase. Calibration was conducted by traversing a dotted calibration board across the measurement volume at regular intervals, similar to what was described in earlier chapters. Note that background noise was removed from the light-field particle images as part of the pre-processing stage, by making use of the minimum pixel value in 100 LF image pairs. DRT-MART approach was used to reconstruct the particle volumes from the light-field particle images, where 10 iterations were adopted. The $36d \times 22d \times 11d$ physical measurement volume was reconstructed into 3200 × 2150 × 200 voxels, before multi-grid cross-correlations were used to obtain the velocity fields.

With an initial and final interrogation volumes of 256 × 128 × 32 voxels and 128 × 64 × 16 voxels respectively, this gave rise to a vector resolution of $0.704d$, $0.352d$ and $0.224d$ in the x, y, and z directions of the measurement volume. Median filtering was utilised to reject spurious vectors and replacement vectors were determined through linear interpolations. GPU accelerated work processes in terms of reconstructing the particles and cross-correlations was carried out by utilising Nvidia GPU cards with CUDA codes.

Figure 6.8 shows a series of velocity field results extracted from the 3D time-averaged velocity fields, where they characterise the synthetic jets studied here. In

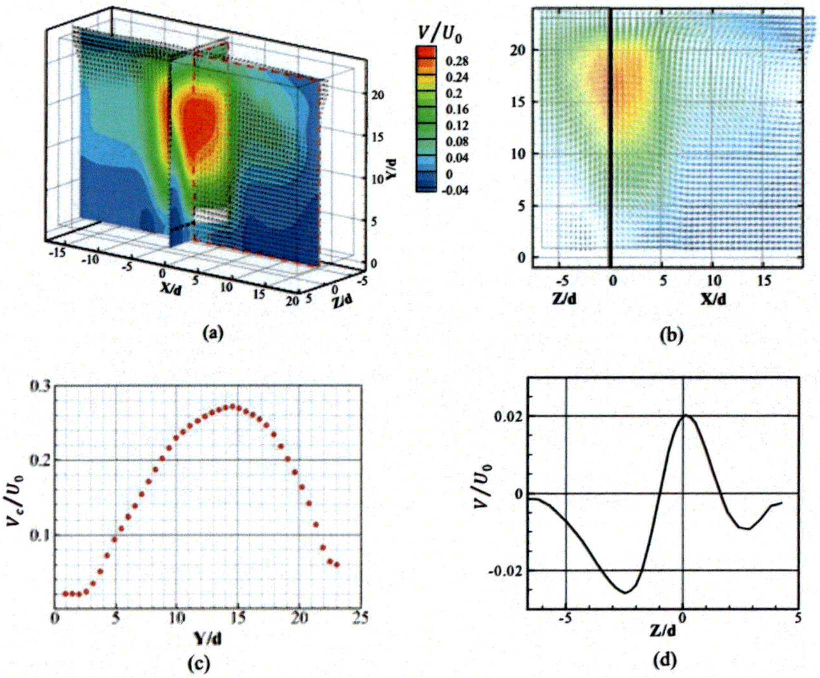

Fig. 6.8 **a** 2D velocity fields extracted from the 3D time-averaged velocity fields, **b** 2D velocity vector fields along $z/d = 0$ and $x/d = 0$ take along xy plane, **c** mean centreline velocity variation, $V_{c(y)}/U_0$ taken along the y direction and **d** axial velocity variation V/U_0 taken along $x/d = 0$, $y/d = 2$ and $z/d = -6.6$ to 4.3

Fig. 6.8a where 2D slices of the mean streamwise velocity components along the two orthogonal planes (i.e. xy and yz planes) are shown, where there is a distinct potential core region that comprises of high velocity fluid from the jet at some downstream location, the results here show that the maximum velocity region occurs at some distance away from the jet exit instead. This is unlike a conventional continuous jet where the potential region starts from the jet exit to some downstream location. The 2D velocity fields can be better appreciated in Fig. 6.8b where the vector fields along $z/d = 0$ and $x/d = 0$ in the xy plane are shown side- by-side, where the unique potential core region can be seen clearly. The centreline velocity variation for the synthetic jet was also plotted and it is shown in Fig. 6.8c. These preceding results demonstrate the zero net-mass flux nature of synthetic jets, where the time-averaged velocity (and hence mass) at the orifice exit should be zero. Last but not least, Fig. 6.8d shows that the synthetic jet flow was not as symmetric as envisioned and was later found to be due to the design of the synthetic jet setup here. Some impact of the asymmetric jet flow upon the resulting flow field will be elaborated when more results are discussed.

Note that the maximum velocity region will be limited by the impingement plate, the latter of which limits the growth of the former within the relatively confined working volume. Furthermore, one interesting phenomenon associated with non-circular jets is the "axis-switching" phenomenon, where the major- and minor-axes of such jets interchange with each other as the jet convects downstream. To look into this further, Fig. 6.9a–c show the isosurfaces associated with the jet half-widths in the 3D measurement volume from different perspectives. As the results show, the present synthetic jet appears to undergo axis-switching behaviour since the jet spread grows rapidly along the minor-plane of the jet orifice, as compared to the much more moderate growth along the jet orifice major-plane. For a closer look at the differences in the jet half-widths between the major- and minor-axes, Fig. 6.9d shows the jet half-width profiles. It can be discerned that the growth rate along the minor-axis is higher before $y/d = 15$ and before the jet impinges upon the plate at $y/d = 24$ location. As a further demonstration on the measurement qualities of LF-PIV, Fig. 6.10 shows the 2D cross-stream slices of the time-averaged impinging synthetic jet flow, as well as the U and V velocity profiles along different axes. In this case, the cross-stream evolutions of the jet potential core can be better discerned and additionally, it should be mentioned that the current LF-PIV measurements were limited by two conflicting needs—dynamic range and particle density. Despite the careful selection of interrogation volumes to satisfy the one-quarter rule for good cross-correlation outcomes, they may still be too big for the typically smaller W velocity component in the z

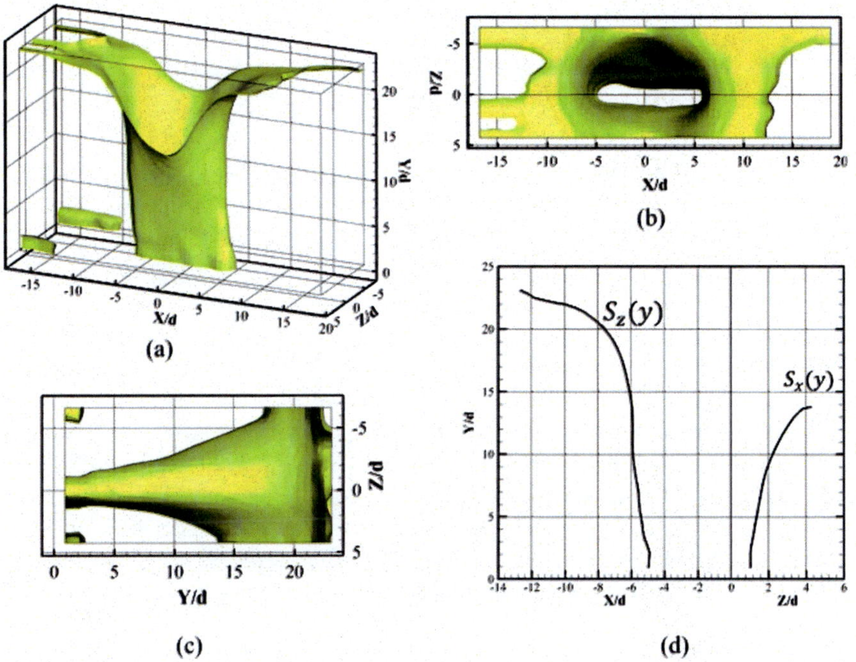

Fig. 6.9 **a** 3D view, **b** top-view and **c** side-view of the 3D isosurfaces associated with the mean jet half-widths. Trajectories of the jet half-width isosurfaces in both x and z directions are shown in greater detail in (**d**)

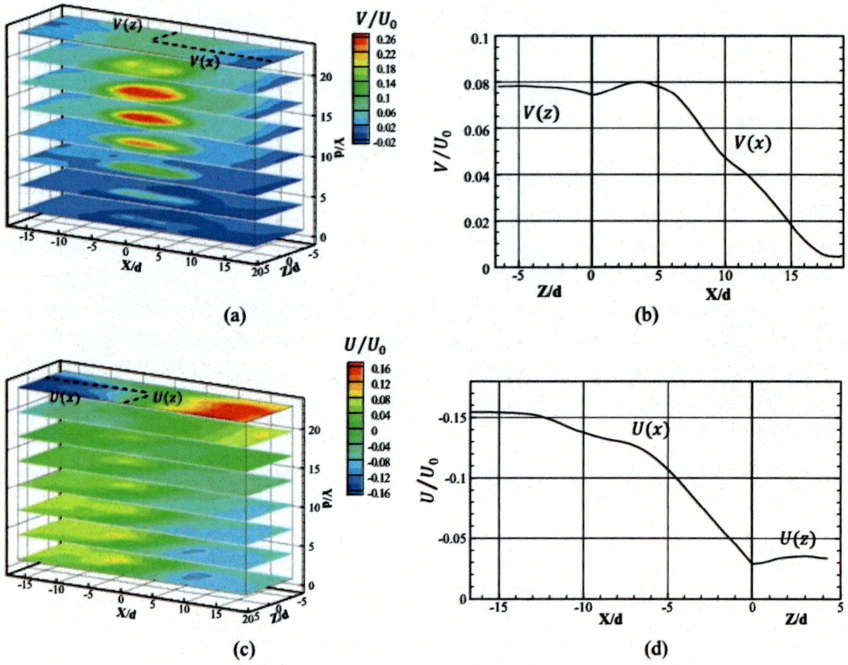

Fig. 6.10 2D cross-stream slices of the 3D time-averaged velocity field results for **a** V/U_0 and **c** U/U_0 velocity components, as well as **b** V/U_0 velocity profiles taken along $x/d = 0$ to 18.7, $y/d = 23$, $z/d = 0$ and **d** U/U_0 velocity profiles taken along $z/d = -16.7$ to 0, $y/d = 23$, $z/d = 0$

direction. If smaller interrogation volumes were to be used to suit the W velocity component, LF-PIV would suffer from the limitations imposed by the MLA resolution, whereby insufficient particles would be available within each interrogation windows for accurate measurements. Due to these concerns, results for W velocity component will be omitted from the discussions here.

Moving away from time-averaged results, Fig. 6.11 shows the first phase-averaged results here where they depict both the velocity vectors and contours of the ejection and suction phases associated with the synthetic jet at various timings. Note that the ejection phases are from $t/\tau = 0/20$–$9/20$ and suction phases are from $t/\tau = 10/20$–$19/20$. Results show that a vortex-ring is formed by $t/\tau = 2/20$ and it travels towards the impingement plate progressively before it collides upon the plate at around $t/\tau = 7/20$. The flow cycle repeats with a clear negative velocity region at the orifice exit at $t/\tau = 15/20$. Due to the alternate blowing and suction phase, a saddle point will be formed at some point between the orifice and the impingement plate and its location varies during the suction phases. Figure 6.12 shows the variation in its location away from the impingement plate, D_{so}, during a typical suction phase, where its locations were determined by identifying locations where all phase-averaged velocity components are all zero. The results agree well with the expectation that the saddle should move gradually closer towards the orifice (i.e. larger D_{so}) during the suction

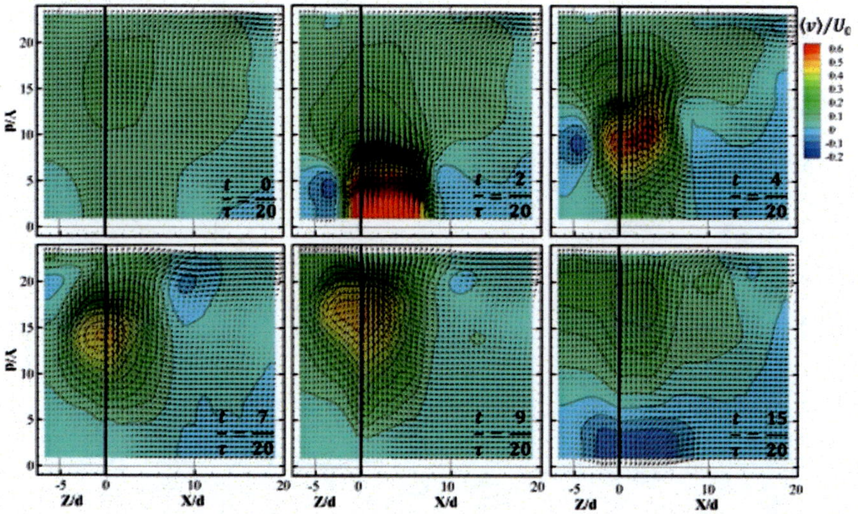

Fig. 6.11 Phase-averaged velocity vector and contours for the synthetic jet production and its impingement upon the plate. Left-hand-side shows results along $x/d = 0$, while right-hand-side shows results along $z/d = 0$

Fig. 6.12 Saddle location variations during the suction phase

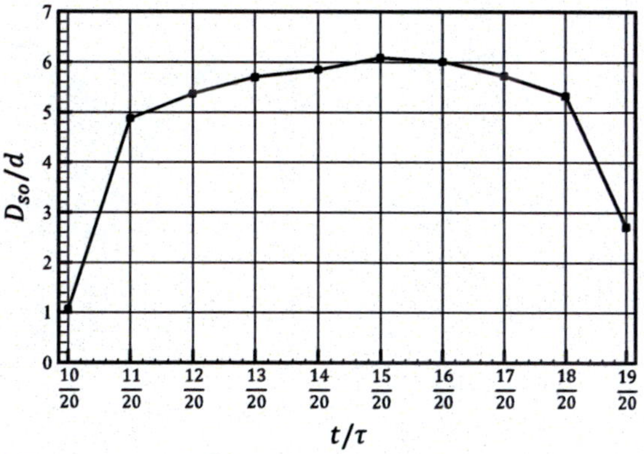

phase. Therefore, it can be seen that the typical behaviour of synthetic jets can be captured successfully by the present LF-PIV setup and procedures.

Next, 3D isosurfaces of the transient flow behaviour associated with the synthetic jet impingement are presented in Fig. 6.13, where the 3D vortex structures were identified based on the velocity gradient tensor $\boldsymbol{D}$, as

$$D \equiv [d_{i,j}] = [v_r \; v_{cr} \; v_{ci}] \begin{bmatrix} \lambda_r & & \\ & \lambda_{cr} & \lambda_{ci} \\ & -\lambda_{ci} & \lambda_{cr} \end{bmatrix} [v_r \; v_{cr} \; v_{ci}]^{-1} \tag{6.11}$$

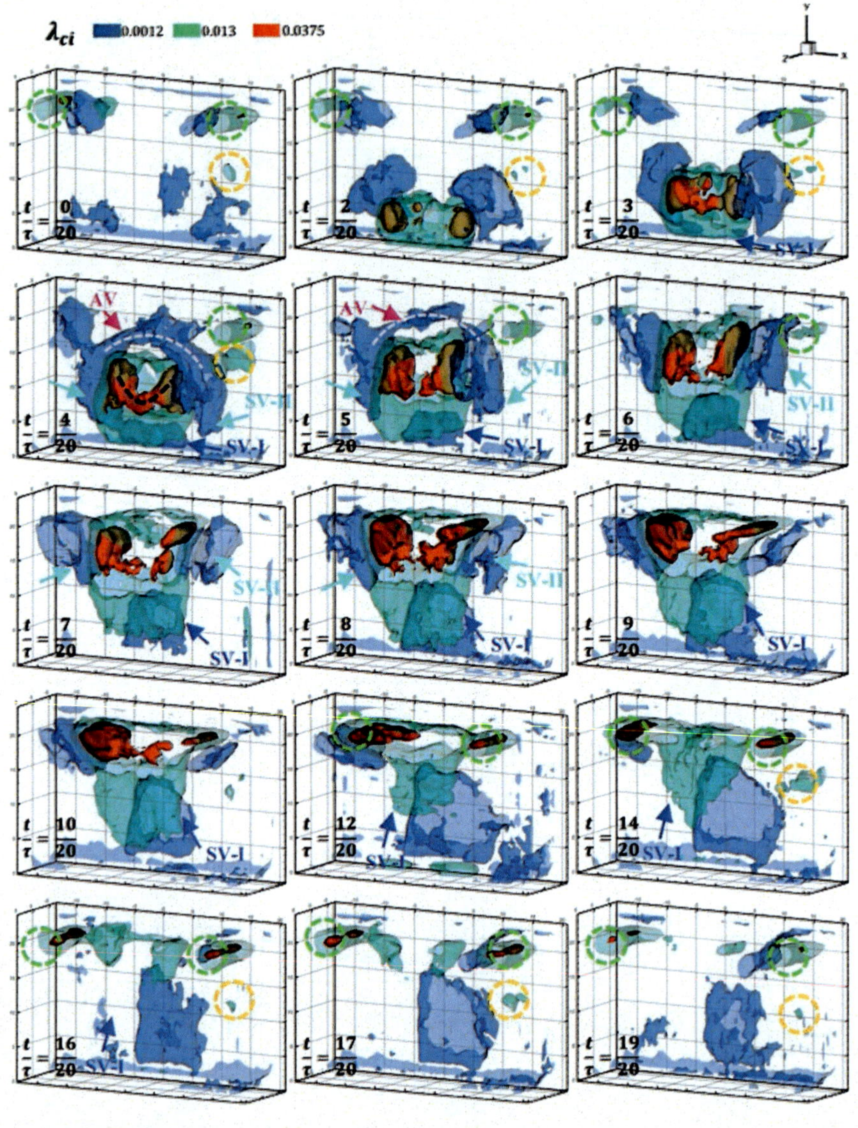

Fig. 6.13 Phase-averaged isosurfaces showing the multitude of vortex structures produced by the present flow scenario, such as the vortex-ring, arc vortex, sweeping vortices, as well as streamwise vortices I and II

As the various key vortex structures differ in terms of their flow strengths, three different λ_{ci} values of 0.0012, 0.013 and 0.0375 were used to identify them. These phase-averaged results are presented in Fig. 6.13 and they show the bending behaviour of the high aspect-ratio vortex-ring at $t/\tau = 4/20$ (see red isosurfaces with dashed line) as part of its axis-switching behaviour at each instance of speaker excitation. As the flow progresses, the vortex-ring structure appears to break up after interacting with the multitude of vortex structures existing in the flow field but continue to collide with the flat plate thereafter. Interestingly, an "arc vortex" of opposite bending direction (see blue isosurfaces with white dashed line) is formed ahead of the vortex-ring structure and actually collides with the flat plate before the vortex-ring structure. Nevertheless, both these vortex structures break up upon impinging upon the flat plate and produce what are termed as "sweeping vortices" (highlighted by green circles). Note that the above vortex structures are by no means the only significant structures that can be observed in the figure. Other relatively coherent structures that can be discerned from the figure include streamwise vortices, known as streamwise vortex I and II (i.e. SV-I and SV-II), that are produced as part of the overall transient flow outcome. A closer look of the preceding behaviour is presented in Fig. 6.14, where it shows the impinging behaviour as viewed normally to the xy- and xz-planes.

In particular, Fig. 6.15 shows how the asymmetric synthetic jet behaviour detected earlier from its characterisation affects the impingement behaviour. It can be observed that the vortex structures are asymmetric and the more dominant vortex structure causes the right-hand-side segment of the primary vortex ring to move faster. Last but not least, Fig. 6.16 shows where the secondary structure within the confine of the vortex-ring structure can be observed, alongside with the sweeping vortex. It is worthwhile to mention that the study revealed that LF-PIV tends to have limited ability in resolving complex structures at this point in time, when compared to other PIV approaches based on similar camera resolution. However, the convenience of a single camera based PIV solution, as well as the expectations that camera technology will continue to advance rapidly, mean that improvements in the measurement fidelity

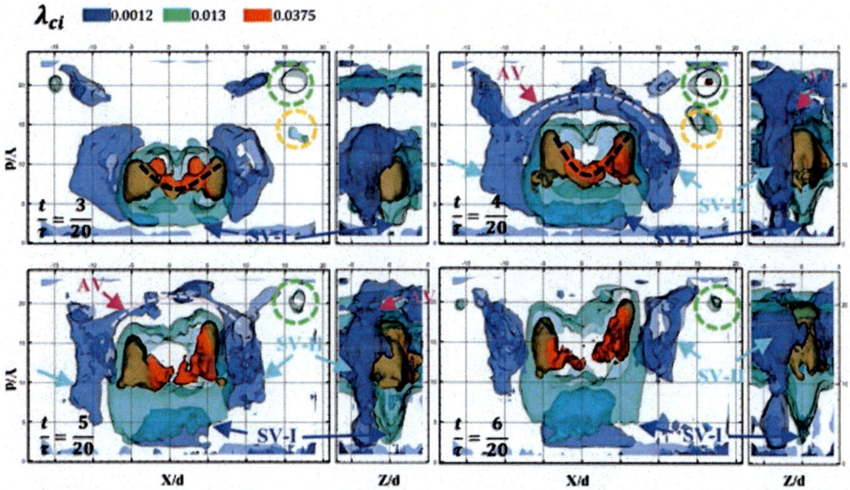

Fig. 6.14 Closer look at the isosurfaces from viewing angles normal to the *xy* and *xz*-planes at various phases

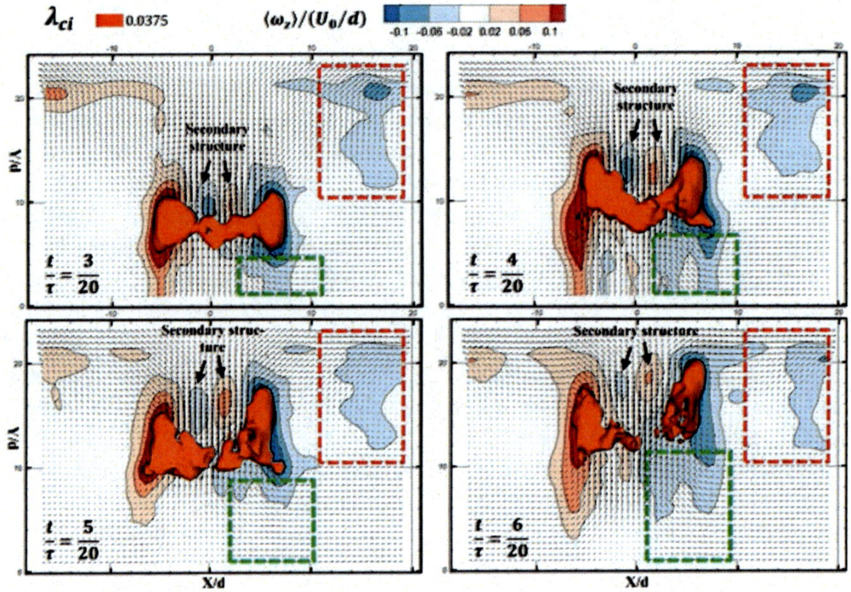

Fig. 6.15 3D isosurfaces and 2D vorticity maps as viewed along the *xy*-plane, showing the secondary structure and sweeping vortex

and temporal resolution will see LF-PIV approach becoming more mature, cost effective and attractive.

Linear Cascades

In the realms of compressor or turbine stage research, measurements of the highly three-dimensional flows in-between the various blades can prove to be, more often than not, very challenging. This can be attributed to a few considerations—firstly, circumferential distribution of the blades about the compressor/turbine shaft makes curved optical access designs that does not introduce distortionary imaging issues challenging. Secondly, optical access for a rotating compressor/turbine setup is limited in the first place due to the geometry and high rotational speeds, the latter of which could lead to high structure stresses that potentially introduce weak spots in the setup. Thirdly, vibrations due to the high-speed rotations may increase the experimental uncertainty levels of any measurements. These are no means the only issues and there are other on top of these as well. To overcome some of these issues and as a matter of simplifying the experimental conditions, as well as to improve the safety and accuracy of conducting experiments on compressor or turbine blades, linear cascades of the said blades are very often utilised instead. As the name implies, a linear cascade is when the compressor or turbine blades are not physically fixed

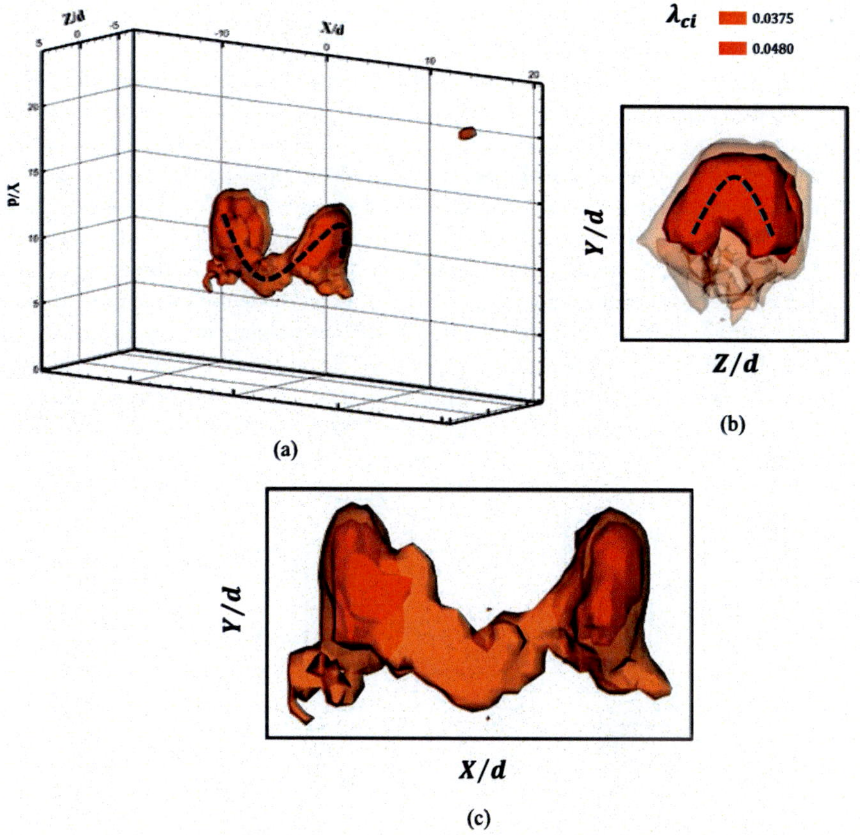

Fig. 6.16 A closer look at the axis-switching behaviour of the vortex ring structure produced by the synthetic jet

around a shaft circumferentially but instead, "unrolled" into a linear configuration where the blades are arranged in a straight row. The blades typically remain stationary with a flow introduced into the inlet and exit from the outlet after passing over the blades. Such a configuration reduces the complexities drastically and prove to be highly useful as a close approximation to the rotating configuration. Nonetheless, it remains challenging and tedious to obtain full-field measurements of the flow fields between the blades, considering the limited spaces and difficulties in achieving good illumination with minimal blockages with the curved blades. While Tomo-PIV approaches appear to be the best strategy, limited optical access for multiple-camera setups in certain configurations may continue to limit the measurement volume that it is able to handle. Hence, most earlier studies continue to make use of conventional 2D or stereoscopic PIV approaches, such investigations by Palafox et al. (2008), Bloxham et al. (2009), Alhaj and Seume 2010), Hu et al. (2014), Pu et al. (2014), Liu et al. (2019) among others. In this case then, the single-camera approach adopted in LF-PIV technique could offer some benefits in terms of simplifying experimental requirements while capturing 3D flow fields.

In light of these potential benefits, single camera LF-PIV was used to investigate the 3D flow behaviour associated with a 11-blade linear cascade (Xu et al. 2019) There are two objectives—firstly, to see if single camera LF-PIV could better facilitate flow measurements within confined spaces, such as the regions between the blades here. Secondly, particular attention was paid towards capturing the corner flow separations and end-wall effects under different blade angles-of-attack (AOA), which would be of significant interest towards optimising blade flows. Figure 6.17 shows the subsonic wind tunnel used for the study, where flow conditioning devices were used to reduce the free-stream turbulence levels during the flow measurements. On the other hand, Fig. 6.18 shows how the linear cascade was set up at the wind tunnel nozzle, where the blade AOA could be adjusted by rotating them directly. There was a total of 11 identical blades located within a 620 mm × 200 mm × 10 mm test section, where the latter had two acrylic covers as its upper and lower surfaces. Blade AOA could

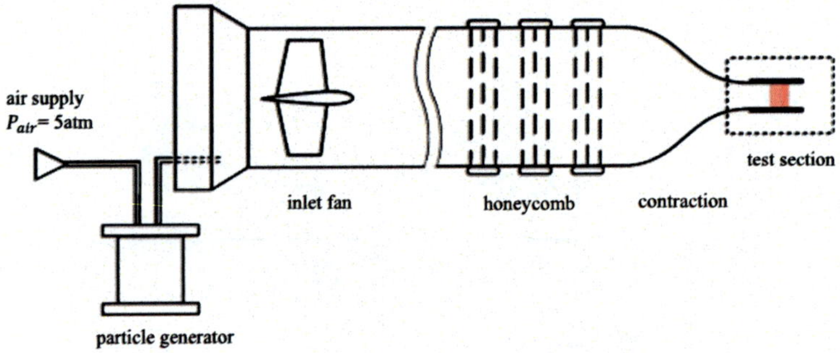

Fig. 6.17 Schematics of experimental setup showing the wind tunnel and the linear cascade located at the wind tunnel nozzle, as well as the particle seeding arrangements

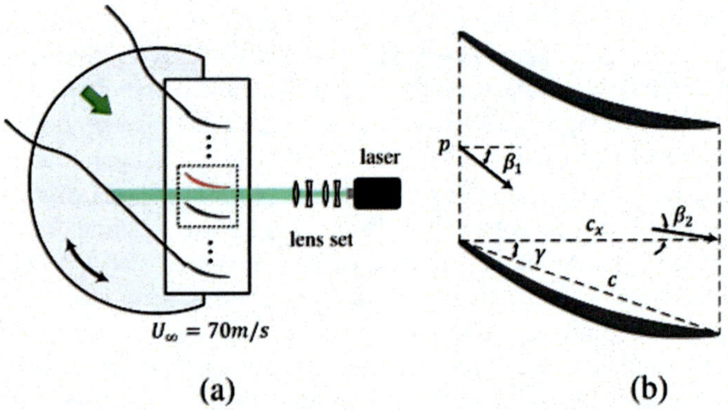

Fig. 6.18 Experimental arrangement for the **a** linear cascade at the wind tunnel nozzle from the top-view and geometrical configuration for the linear cascade blades

Table 6.1 Geometrical details of the linear cascade

Chord, c (mm)	70.4
Axial chord, c_x (mm)	66.2
Pitch, p (mm)	52.3
Span, h (mm)	100
Inlet design flow angle, β_1 (°)	38
Outlet design flow angle, β_2 (°)	1.5
Pressure side tangent stagger angle (°)	20
Blade number	11

be adjusted by rotating the circular platform upon which the blades are attached and Table 6.1 lays out the geometrical details of the linear cascade. The free-stream velocity used for the present study was set to 70 m/s and the Reynolds number based on the blade chord length, c, was approximately $\mathrm{Re}_c = 3.6 \times 10^5$. As the Mach number is about 0.21, the flow was assumed to be entirely incompressible.

LF-PIV experiments were carried out using an in-house modified light-field Imperx B6640M based camera, similar to the one utilised by Shi et al. (2016). The LF-PIV and linear cascade setups are shown in Fig. 6.19 for a better appreciation of the experimental arrangements. Due to safety concerns, atomised Bis(2-ethylhexyl) sebacate was introduced into the wind tunnel via its inlet using a particle generator

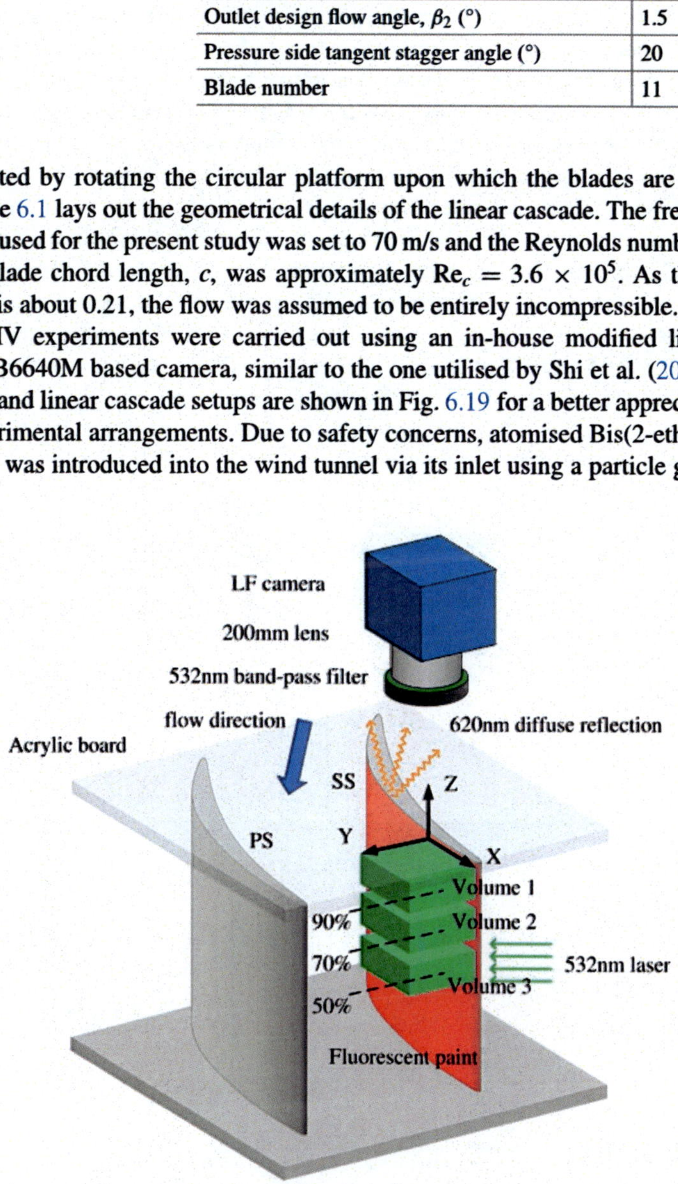

Fig. 6.19 Schematics of the linear cascade experimental and LF-PIV setup

at 5 times the atmospheric pressure as seeding particles with a nominal size of about 1 μm. During the course of the experiments, it was observed that non-uniform distribution of seeding particles impacted negatively upon the capturing of satisfactory images for post-processing. Note that LF-PIV is highly sensitive towards particle distribution and good distribution is key towards satisfactory measurement results. Nevertheless, 150 valid particle image-pairs were captured for each test conditions. It should be highlighted that, at this moment and at least for this study, LF-PIV faced similar issues as with Tomo-PIV in terms of covering significant measurement volume size. Hence, three separate measurement volumes were captured as shown in Fig. 6.19, so as to cover a larger overall volume of investigation (albeit with small 3 mm gaps between adjacent measurement volumes). Having said that however, the ability to cover larger measurement volumes should improve with camera resolution and technology, the latter of which is advancing rapidly. With three separate measurement volumes, the plane of focus was centred at three different locations at $z/h = 0.5$, 0.7 and 0.9. The overall volume of investigation was 390 mm × 260 mm × 170 mm, where it was divided into 2200 × 1466 × 320 voxels. Initial and final interrogation volumes used for cross-correlations with 75% overlapping ratio were 128 × 128 × 64 voxels and 64 × 64 × 32 voxels respectively, with a vector resolution of 1.16 mm × 1.16 mm × 1.93 mm attained.

For a quick appreciation of the flow fields that were encountered here, Fig. 6.20 shows the instantaneous streamwise velocity distributions at four different spanwise

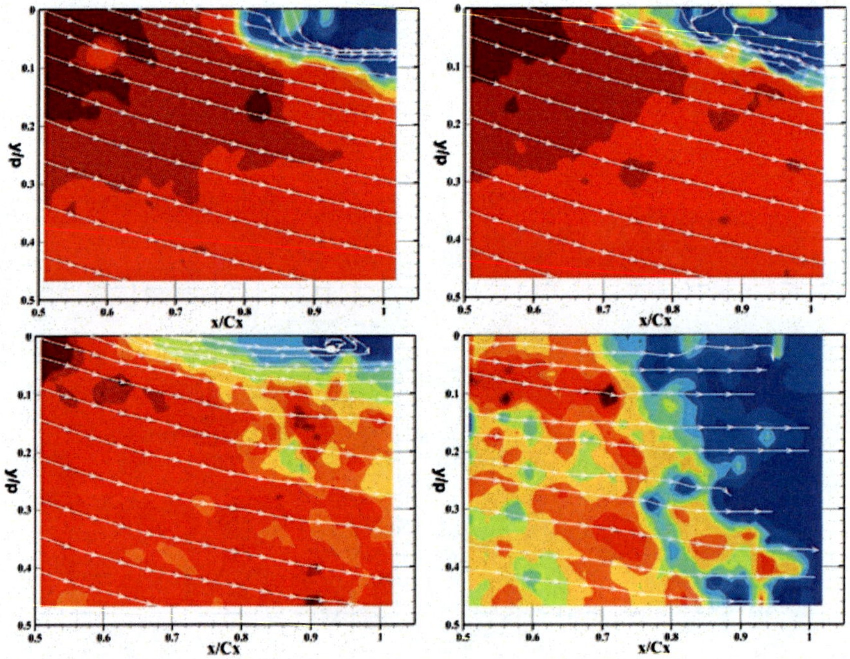

Fig. 6.20 Instantaneous streamwise velocity distributions at different spanwise locations close to the blade trailing-edge

locations close to the blade trailing-edge, namely at z/h = 82.0, 86.1, 91.0 and 95.9% from the lower acrylic board surface (see Fig. 6.19). It can be seen that as the measurement plane moves progressively closer to the top acrylic board, increasingly larger regions of low velocity magnitudes are observed. This is due to corner effects caused by the intersection of the blade and upper acrylic board becoming increasingly more prominent as the measurement plane moves closer to that area. For an even closer look and to assess just how persistent and dominant the corner effects/separations are, mean cross-stream speed, turbulent kinetic energy (TKE) and vorticity distributions are presented in Figs. 6.21, 6.22 and 6.23. Figure 6.21 show the presence of two separate low-speed regions—one associated with the boundary layer along the suction-side of the blade and another due to the corner effects as mentioned earlier on. The boundary layer grows progressively thicker with corresponding increase in near-wall low speed region size and the same can be said for the slower corner flows. Interestingly, a low speed region can also be observed along the top and that could be due to the boundary layer along the lower surface of the top acrylic board. The existence and subsequent behaviour of these low speed regions have significant impact upon the resulting TKE and vorticity distributions as well, as can be observed from Figs. 6.22 and 6.23. For a better idea on how the distribution of low speed regions is for the blade, Fig. 6.24 shows the isosurfaces plotted with a cut-off speed of 30 m/s. For more efficient blade designs and working conditions, these results are able to identify flow issues that should be mitigated for optimal blade flow fields.

These observations are comparable with past studies and show that LF-PIV approach is able to resolve the key flow behaviour associated with a blade from a linear cascade. While issues surrounding non-uniform particle density and limited measurement volume should be resolved before one expects LF-PIV to reach greater measurement resolution without the need to combine several smaller measurement volumes, they are not entirely dissimilar to those faced by other PIV approaches. On the other hand, the convenience and ease of using only a single camera reduces the complexity of at least one particular aspect associated with such studies.

Supersonic Jets

Single camera LF-PIV approach has been adopted by Ding et al. (2018), Sheikh (2019) to investigate Ma = 1.3 supersonic jet flows under a variety of nozzle-pressure ratio (NPR) conditions, where it ranged between NPR = 2.2–3.2. Note that perfect expansion occurred at NPR = 3.1 and thus, the test cases included both over- and under-expanded conditions. Due to the significant flow velocities which produce a multitude of shock structures, one of the major concerns was whether the LF-PIV approach would be sufficiently robust to capture and resolve them satisfactorily. Conventional Tomo-PIV measurements on supersonic flows have traditionally been

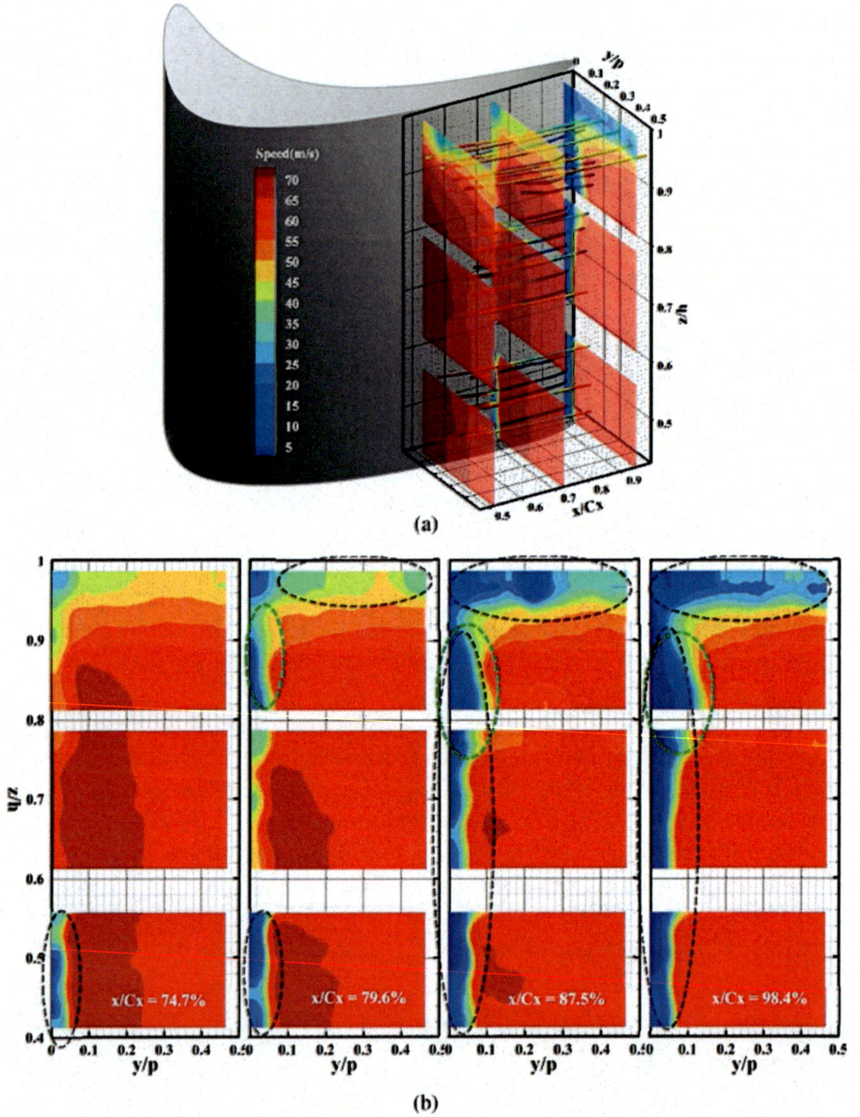

Fig. 6.21 Mean cross-stream speed magnitude distributions at $x/c_x = 74.7, 79.6, 87.5$ and 98.4%

highly challenging, even when multiple cameras are available. For single camera LF-PIV, the technical challenges will be even greater, due to higher particle and illumination requirements, among others. To the best of the authors' knowledge, these studies were some of the first to be conducted on supersonic flows. Hence, special attention was paid towards the ability of LF-PIV measurements to resolve

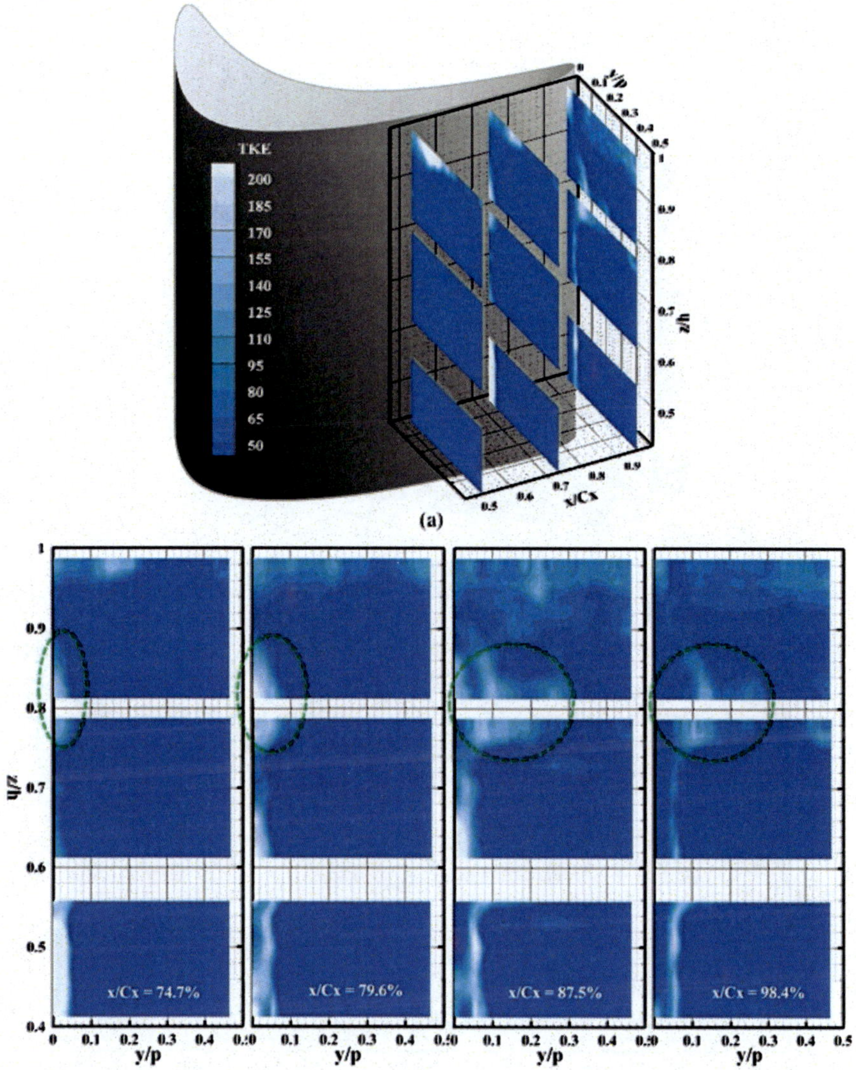

(a)

(b)

Fig. 6.22 Mean cross-stream turbulent kinetic energy distributions at $x/c_x = 74.7$, 79.6, 87.5 and 98.4%

the shock structures and the velocity fields between the sharp, discontinuous shocks properly and adequately during the study.

Highly compressed air was produced by an air compressor, which passed through a series of settling tanks, air dryer and filters before been directed into the jet apparatus, as shown in Fig. 6.25. The aluminium jet nozzle and throat inner diameters were 10 mm (D) and 9.22 mm respectively, as shown in the photos of the setup and

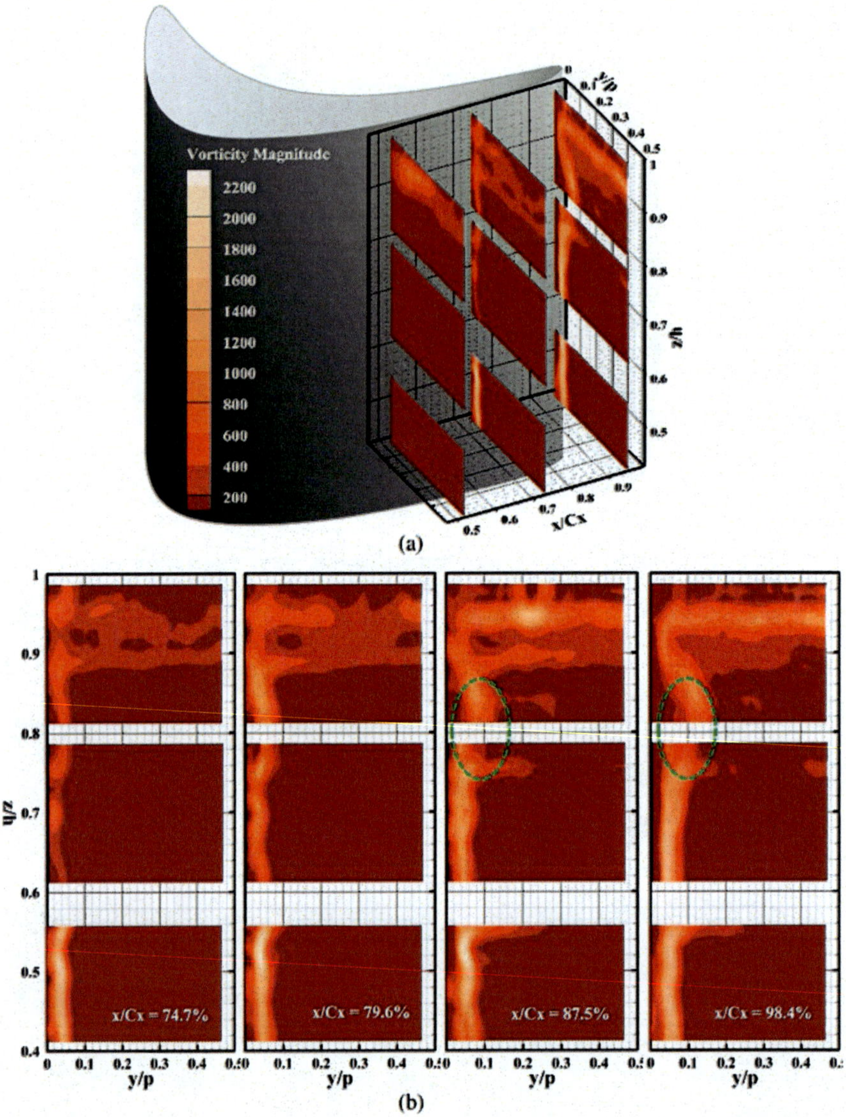

Fig. 6.23 Mean cross-stream vorticity distributions at $x/c_x = 74.7, 79.6, 87.5$ and 98.4%

nozzle design schematics presented in Fig. 6.26. 1 μm nominal diameter aluminium oxide particles were produced by a particle generator and introduced into the flow delivery line upstream of the jet apparatus, for the purposes of seeding the supersonic jets during the experiments. The supersonic jets were then exhausted into an octagonal-shaped Plexiglas tank, where an exhaustion system was available to purge the seeded air. Before LF-PIV experiments were conducted, Schlieren visualisations

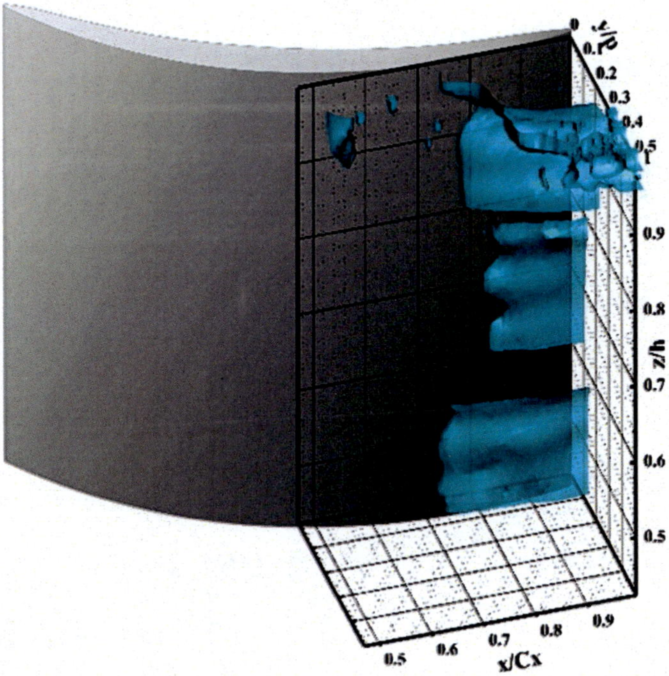

Fig. 6.24 Isosurfaces of low-speed regions for the present blade

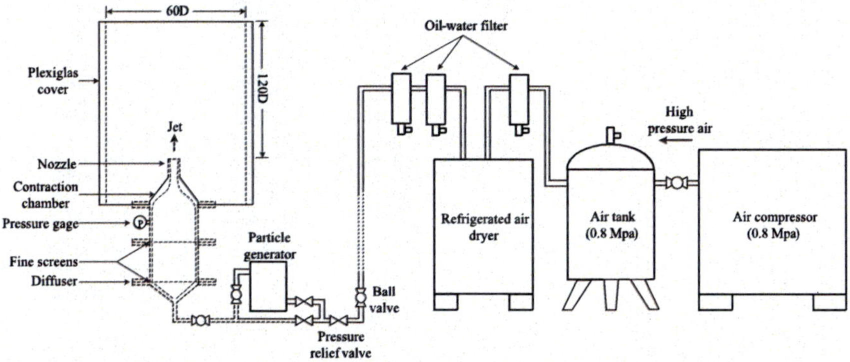

Fig. 6.25 Experimental setup used for LF-PIV measurements of supersonic jet flows

and conventional 2D-PIV experiments were first carried out to get first-hand appreciation of the supersonic jet flows, as well as to establish some benchmark results that could be used to compare with the LF-PIV results.

For Schlieren visualisations, a modified Z-type Schlieren system shown in Fig. 6.27 was used to capture qualitative images of the resulting shock structures produced by the supersonic jet. Such a system had been used successfully by Wu

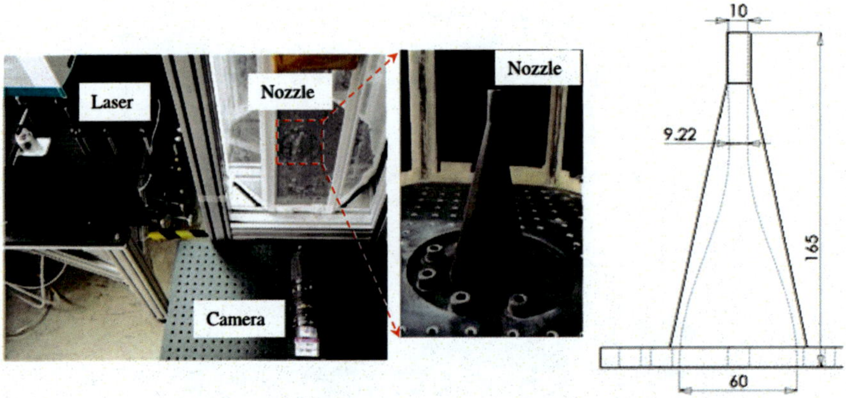

(a) (b)

Fig. 6.26 Photos of the experimental setup and nozzle design schematics

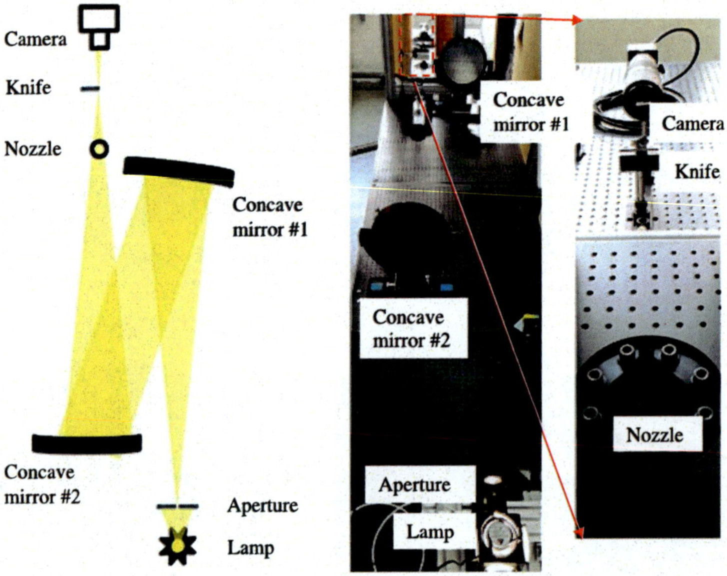

(a) (b)

Fig. 6.27 2D schlieren setup used to image the shock structures of the supersonic jet flows

and New (2017), Lim et al. (2018), Zang et al. (2018), Mariani et al. (2019, 2020) and Wu et al. (2019). As the figure shows, the light source originated from a 200 W broadband light LED lamp and the light beam passed through a pinhole aperture, before it was reflected off two 300 mm diameter parabolic mirrors. A knife edge was used and oriented parallel to the jet flow axis. For 2D-PIV measurements, laser beam from a 500 mJ/pulse dual-head Nd:YAG laser was formed into a ~ 1 mm thin laser sheet using cylindrical convex and concave lenses and directed into the measurement area as shown in Fig. 6.28a. The measurement window was located at approximately 0.22D above the nozzle and had a physical size of about 1.4D × 2D. A 2048 × 2048 pixels, Imperx B2041 PIV camera was then used to capture the particle image-pairs. Conventional multi-grid cross-correlation particle-image post-processing was used, where initial and final correlation windows of 64 × 32 pixels and 32 × 16 pixels were used, together with 75% overlap in both directions. As for LF-PIV, the setup was generally similar with the exception that a 6600 × 4400 pixels in-house LF camera based on Imperx B6640M PIV camera was used, and that different optics were used to produce the volumetric laser illumination. As shown in Fig. 6.28b, this demonstrates the "plug and play" convenience of the single camera LF-PIV approach under certain conditions as one of its biggest advantages.

The measurement volume was located at approximately 0.39D above the nozzle and covered a physical volume of 1.25D × 1.25D × 1.25D. The same laser was used as the illumination source but in this case, a volumetric illumination was produced for

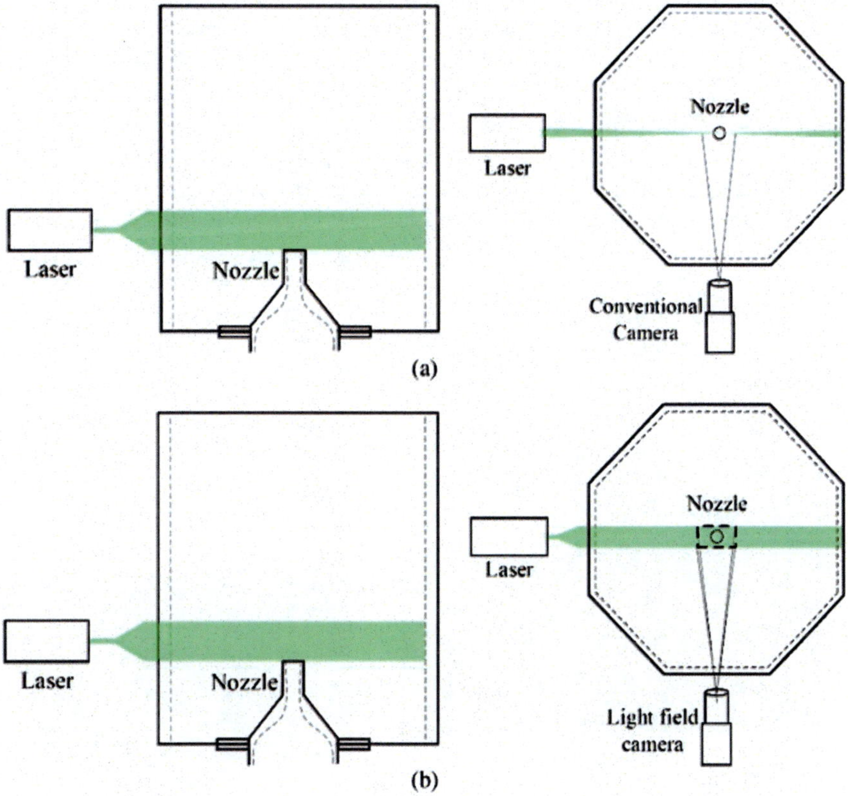

Fig. 6.28 Schematics of the experimental setups used for 2D- and LF-PIV measurements. Note the high level of similarity between the two with the exception of the camera type and illumination (i.e. sheet or volumetric)

LF-PIV instead, where the latter was formed by a combination of two sets of cylindrical convex and concave lenses. To reconstruct the captured LF particle images, DRT-MART technique and optical corrections (Shi et al. 2017, 2018) were used. The pixel voxel ratios (PVR) used for in the x, y and z directions were 3, 3 and 10 respectively, which produced a total number of 800 × 800 × 240 voxels for the reconstruction domain. This led to a reconstruction resolution of 0.0165 mm × 0.0165 mm × 0.055 mm per voxel. For subsequent post-processing to arrive at the velocity fields, 3D multi-grid cross-correlation with initial and final interrogation volumetric sizes of 128 × 128 × 64 pixels and 64 × 64 × 32 pixels were used respectively. Note that these post-processing stages were enhanced through the use of GPU acceleration via Nvidia CUDA codes.

As this was one of the first time that LF-PIV was used on a supersonic flow scenario, an example of the captured particle-images by the light-field camera for the supersonic jet operating at NPR = 2.6 overexpanded condition before post-processing is shown in Fig. 6.29. One initial concern was that the presence of the shock structures could smear or distort the particle images adversely and affect the reconstruction quality. However, the light-field particle image appeared to be relatively free of any visible distortions or smearing. As the procedures were similar regardless of the exact NPR used, only results obtained for this NPR will be presented here for the sake of brevity. 100 sets of 2D and light-field particle images were subsequently post-processed and averaged to arrive at the mean velocity distributions for the sake of comparisons between each other, as well as with typical flow features expected for a supersonic jet issuing under approximately similar conditions.

Figure 6.30 shows the 2D-PIV mean streamwise, radial velocity components and Schlieren image captured for the supersonic jet, where comparisons with earlier studies show that the present 2D-PIV measurements had successfully captured the velocity fields despite the presence of abrupt shock structures within the jet potential core. Note that the red and green boxes indicate the 2D-PIV and LF-PIV measurement

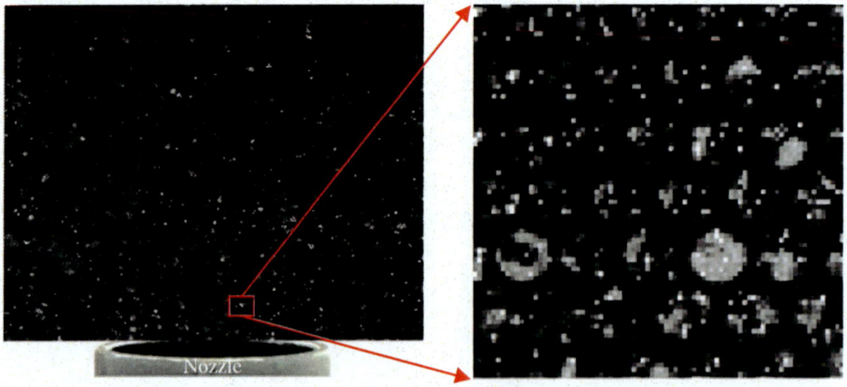

Fig. 6.29 Example of a particle-image captured by the light-field camera and a close-up view of a selected region within the supersonic jet potential core

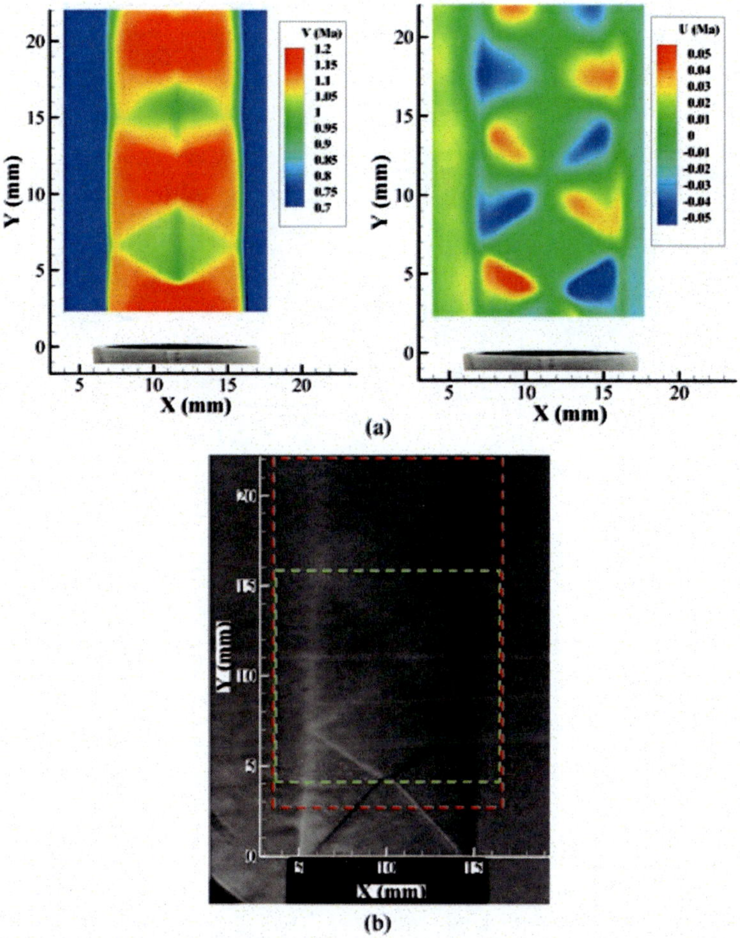

Fig. 6.30 **a** Mean streamwise (V) and radial (U) velocity distributions captured by 2D-PIV, as well as the shock structures visualised using Schlieren imaging, within the supersonic jet flow. Note that the red and green sections indicated in (**c**) are the measurement regions for the 2D- and LF-PIV approaches respectively

regions. The shock structures associated with the supersonic jet at the present working conditions resemble closely to those observed under relatively similar conditions by previously mentioned studies, as revealed by their qualitative 2D Schlieren, quantitative 3D Schlieren and numerical simulation results. The absolute Mach number distributions are also in satisfactory agreement with the expectations of an NPR = 2.6 supersonic jet. On the other hand, Fig. 6.31 shows 2D results extracted from LF-PIV measurements and while the extent of the measurement region in the streamwise direction is relatively shorter than that for the 2D-PIV, the results are comparable between the two different PIV techniques. For instance, abrupt changes to the flow velocity caused by the formation of oblique shocks at the nozzle trailing-edge and

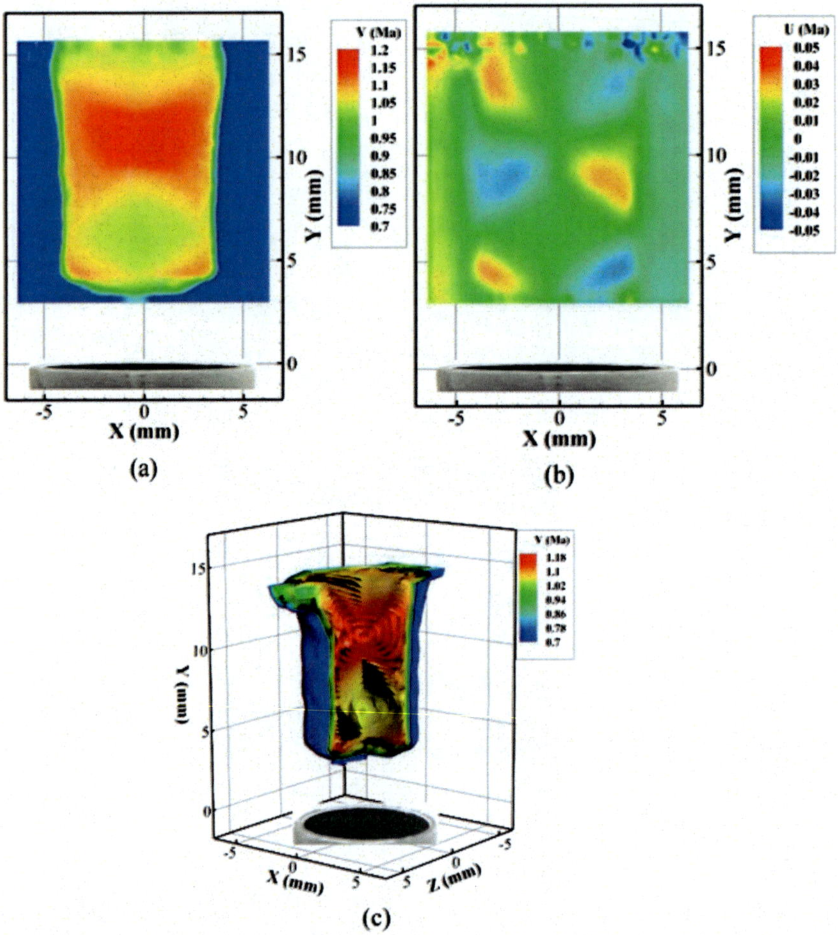

Fig. 6.31 **a** Mean streamwise (V) and radial (U) velocity distributions, as well as reconstructed 3D velocity isosurfaces based on all three velocity components captured by LF-PIV

their intersecting with each other an approximately half a diameter from the nozzle exit are captured by both approaches and demonstrate that it is feasible for LF-PIV to be implemented for highly compressible flows. Furthermore, the presence of shock structures did not appear to introduce gross inaccuracies or distortions. Having said that, it should be reminded that 2D-PIV results remain discernibly "cleaner" than LF-PIV ones with lower noise levels. Additionally, the velocity changes registered in the 2D-PIV results are more abrupt than those depicted in LF-PIV, and more representative of the sharp velocity variations one expected to see in supersonic potential cores where shock structures are present. This is likely due to the presence of some 3D optical distortions to the light-field particle images caused by strong compressibility effects and should be correctable through additional calibration procedures.

Nonetheless, what LF-PIV approach is able to provide over 2D-PIV much more easily in this case can be appreciated in Fig. 6.25c, where 3D velocity isosurfaces of the shock structures are shown. While there remain some technical refinements to be studied and made, in particular improvements to the overall measurement resolution and ability to resolve high velocity gradient regions in compressible flows, this study demonstrates that LF-PIV implementation on supersonic flows is feasible and may eventually prove to be a fruitful one.

References

Alhaj O, Seume JR (2010) Optical investigation of profile losses in a linear turbine cascade

Atkinson C, Buchner AJ, Kitsios V, Soria J (2015) A multi-camera PIV system for the study of self-similar APG turbulent boundary layers. 11th Int Sym Part Image Velocim Sept 14–16

Atkinson C, Buchner AJ, Sekimoto A, et al (2016) Experimental measurements of a self-similar adverse pressure gradient turbulent boundary layer. In: Proceedings of the 20th Australasian fluid mechanics conference, Perth, Australia, Australasian Fluid Mechanics Society

Berk T, Gomit G, Ganapathisubramani B (2016) Vectoring of parallel synthetic jets: a parametric study. J Fluid Mech 804:467–489

Bloxham M, Reimann D, Crapo K, et al (2009) Synchronizing separation flow control with unsteady wakes in a low-pressure turbine cascade

Buxton ORH, Ganapathisubramani B (2010) Amplification of enstrophy in the far field of an axisymmetric turbulent jet. J Fluid Mech 651:483–502

Cater JE, Soria J (2002) The evolution of round zero-net-mass-flux jets. J Fluid Mech 472:167–200

Deem EA, Zhang Y, Cattafesta LN et al (2016) On the resolution of plenoptic PIV. Meas Sci Technol 27:84003

Ding J, Lim D, Sheikh S et al (2018) Volumetric measurement of a supersonic jet with single-camera light-field PIV. In: The 18th international symposium on the application of laser and imaging techniques to fluid mechanics, Lisbon, Portugal

Edgington-Mitchell D, Honnery DR, Soria J (2014a) The underexpanded jet Mach disk and its associated shear layer. Phys Fluids 26:1578

Edgington-Mitchell D, Oberleithner K, Honnery DR, Soria J (2014b) Coherent structure and sound production in the helical mode of a screeching axisymmetric jet. J Fluid Mech 748:822–847

Ganapathisubramani B, Longmire EK, Marusic I (2002) Investigation of three dimensionality in the near field of a round jet using stereo PIV. J Turbul 3:16

Ganapathisubramani B, Lakshminarasimhan K, Clemens NT (2007) Determination of complete velocity gradient tensor by using cinematographic stereoscopic PIV in a turbulent jet. Exp Fluids 42:923–939

Ganapathisubramani B, Lakshminarasimhan K, Clemens NT (2008) Investigation of three-dimensional structure of fine scales in a turbulent jet by using cinematographic stereoscopic particle image velocimetry. J Fluid Mech 598:141–175

Hu J, Kong X, Li Z et al (2014) Experimental investigation of aerodynamic interaction between tip leakage flow and spontaneous tip injection flow using 2D-PIV. Exp Therm Fluid Sci 54:127–135

Le Clainche S, Vega JM, Soria J (2017) Higher order dynamic mode decomposition of noisy experimental data: the flow structure of a zero-net-mass-flux jet. Exp Therm Fluid Sci 88:336–353

Lim TT, New TH, Luo SC (2006) Scaling of trajectories of elliptic jets in crossflow. AIAA J 44:3157–3160

Lim HD, New TH, Mariani R, Cui YD (2018) Visual hull based 3D reconstruction of shocks in under-expanded supersonic bevelled jets. Exp Therm Fluid Sci 99:458–473

Liu Q, Zhong S, Li L (2019) Effects of bio-inspired micro-scale surface patterns on the profile losses in a linear cascade. J Turbomach 141:121006
Long J, New TH (2015) A DPIV study on the effects of separation distance upon the vortical behaviour of jet–cylinder impingements. Exp Fluids 56:1–21
Long J, New TH (2016) Vortex dynamics and wall shear stress behaviour associated with an elliptic jet impinging upon a flat plate. Exp Fluids 57:1–18
Long J, New TH (2019) Vortical structures and behaviour of an elliptic jet impinging upon a convex cylinder. Exp Therm Fluid Sci 100:292–310
Mariani R, Zang B, Lim HD et al (2019) A comparative study on the use of calibrated and rainbow schlieren techniques in axisymmetric supersonic jets. Flow Meas Instrum 66:218–228
Mariani R, Lim HD, Zang B et al (2020) On the application of non-standard rainbow schlieren technique upon supersonic jets. J vis 23:383–393
New TH (2009) An experimental study on jets issuing from elliptic inclined nozzles. Exp Fluids 46:1139–1157
New TH, Long J (2015) Dynamics of laminar circular jet impingement upon convex cylinders. Phys Fluids 27:24109
New TH, Tsovolos D (2009) Influence of nozzle sharpness on the flow fields of V-notched nozzle jets. Phys Fluids 21:84107
New TH, Tsovolos D (2011) On the vortical structures and behaviour of inclined elliptic jets. Eur J Mech 30:437–450
New TH, Tsovolos D (2012) On the flow characteristics of minor-plane inclined elliptic jets. Exp Therm Fluid Sci 38:94–106
Palafox P, Oldfield MLG, LaGraff JE, Jones T V (2008) PIV maps of tip leakage and secondary flow fields on a low-speed turbine blade cascade with moving end wall
Pu J, Yu J, Wang J et al (2014) An experimental investigation of secondary flow characteristics in a linear turbine cascade with upstream converging slot-holes using TR-PIV. Exp Therm Fluid Sci 59:56–71
Sheikh S (2019) Application of light field particle image velocimetry in supersonic jet flow. Shanghai Jiao Tong University
Shi S, New TH (2013) Some observations in the vortex-turning behaviour of noncircular inclined jets. Exp Fluids 54:1–11
Shi S, Wang J, Ding J et al (2016) Parametric study on light field volumetric particle image velocimetry. Flow Meas Instrum 49:70–88
Shi S, Ding J, New TH, Soria J (2017) Light-field camera-based 3D volumetric particle image velocimetry with dense ray tracing reconstruction technique. Exp Fluids. https://doi.org/10.1007/s00348-017-2365-3
Shi S, Ding J, Atkinson C et al (2018) A detailed comparison of single-camera light-field PIV and tomographic PIV. Exp Fluids. https://doi.org/10.1007/s00348-018-2500-9
Shi S, Ding J, New TH et al (2019) Volumetric calibration enhancements for single-camera light-field PIV. Exp Fluids. https://doi.org/10.1007/s00348-018-2670-5
Toh K, Honnery D, Soria J (2010) Axial plus tangential entry swirling jet. Exp Fluids 48:309–325
Wu J, Lim HD, Wei X et al (2019) Flow characterization of supersonic jets issuing from double-beveled nozzles. J Fluids Eng 141
Wu J, New TH (2017) An investigation on supersonic bevelled nozzle jets. Aerosp Sci Technol 63:278–293
Xu S, Mei D, Ding J, et al (2019) Linear cascade 3D flow measurement with single-camera light-field PIV. In: The 13th international symposium on particle image velocimetry. München, Germany
Zang B, New TH (2015) On the wake-like vortical arrangement and behaviour associated with twin jets in close proximity. Exp Therm Fluid Sci 69:127–140
Zang B, New TH (2017) Near-field dynamics of parallel twin jets in cross-flow. Phys Fluids 29:35103
Zang B, Vevek US, Lim HD et al (2018) An assessment of OpenFOAM solver on RANS simulations of round supersonic free jets. J Comput Sci 28:18–31

Zhao Z, Buchner A-J, Atkinson C et al (2019) Volumetric measurements of a self-similar adverse pressure gradient turbulent boundary layer using single-camera light-field particle image velocimetry. Exp Fluids 60:1–14

Zhao Z, Ding J, Shi S et al (2021) Volumetric flow characterisation of a rectangular orifice impinging synthetic jet with single-camera light-field PIV. Exp Therm Fluid Sci 123:110327

Chapter 7
Future Developments of Light-field-Based Measurements

T. H. New and Shengxian Shi

Abstract As light-field camera technology and post-processing algorithms continue to improve, and increasingly better measurement resolutions and accuracy levels are realised, the range of new light-field-based measurement technologies is expected to widen in the coming years. This chapter will briefly cover several new emerging measurement technologies and techniques in which light-field-based approaches may be or have been recently used to improve their measurement procedures and fidelity, as well as the level of adoption. Whilst by no means meant to be an exhaustive coverage of all possible measurement scenarios, the purpose of this chapter is to provide insights into how light-field-based approaches may help to accelerate or ease the measurement data acquisition process, as well as opening up new measurement possibilities.

Keywords Light-field · Multispectral radiation thermometry · 3D geometry measurement · Droplet and splash measurement · 3D flame structure reconstruction

Light-field Multi-spectral Radiation Thermometry

Accurate high temperature measurements are critical for applications such as metal manufacturing, gas turbine blade cooling, and rocket engine nozzles (Wen and Mudawar 2004; Xing et al. 2017; Li et al. 2019a). Although thermistors and thermocouples have been widely used in temperature measurements, such intrusive point-wise methods may alter the measurement surface physically (i.e. leading to disturbed flow field near a turbine blade, for example), or are unsuitable in applications where the measurand is fast-moving and under extremely high temperature. As contactless measurement techniques go, radiation thermometry such as spectral,

T. H. New (✉)
School of Mechanical and Aerospace Engineering, Nanyang Technological University, Singapore 639798, Singapore
e-mail: DTHNEW@ntu.edu.sg

S. Shi
School of Mechanical Engineering, Shanghai Jiao Tong University, Shanghai 200240, China
e-mail: kirinshi@sjtu.edu.cn

S. Shi and T. H. New (eds.), *Development and Application of Light-Field Cameras in Fluid Measurements*, https://doi.org/10.1007/978-3-031-01779-7_7

dual-wavelength and multi-spectral thermometry provide promising alternatives for remote high temperature measurements (Lü et al. 2016; Meng and Diebold 2016; Araújo 2017; Liu and Zheng 2020). Amongst these methods, multi-spectral radiation thermometry (MRT) that infers temperature from radiations emitted by more than three wavelengths, is usually preferred for its higher measurement accuracy level (Ackerman 1962; Gardner et al. 1982).

A key issue in MRT however, is how to effectively separate different radiation wavelengths. If the temperature of a static target is desired, multiple radiation wavelengths could be time-multiplexed onto a single detector by a rotary filter wheel or scanning spectrometer. For fast-moving object temperature measurements however, the radiation wavelengths could be separated using a prism, optical grating, partially reflecting mirror or filter systems. These approaches are not without their drawbacks though. Take for instance, temperature control is normally necessary for a prism to achieve wavelength stability, whereas an optical grating requires order sorting that may result in strong polarisation differences between the filtered wavelengths (Cashdollar 1979). Additionally, partially reflecting mirrors, which are typically applied as beam splitters, may introduce significant discrepancies in the spectral channel response (Lu and Yan 2006; Yan et al. 2020). Last but not least, filter systems are normally applied in conjunction with collimating optics, where each detector only captures a portion of the radiation of the selected wavelength.

Note that the above MRT methods relate to point-wise pyrometers that could only detect a handful of discrete temperature data with short time intervals. On the other hand, surface area MRT measurements are currently performed using colour CCD or CMOS imaging sensors, where three radiation wavelengths can be simultaneously captured by red–green–blue (RGB) Bayer filters (Tsyba et al. 2003; Fu et al. 2006). However, measurement accuracy is significantly constrained by the limited number of spectrums, and more wavelengths are necessary to mitigate the effect of unknown spectral emissivity (Fu et al. 2016).

How to fundamentally increase measurement accuracy by capturing wider radiation spectrums whilst keeping the optical system simple, is one of the greatest challenges in MRT and light-field-based imaging could provide a promising solution. It is known that a light-field camera can capture information on the directions and intensities of incident rays simultaneously, and multi-view images can be calculated from one single light-field raw image. If a filter array can be properly designed and correctly coupled with the MLA, it is possible that the incident ray directional information could be traded for the incident radiation spectral information. As such, multi-spectral images could be generated from just one raw light-field radiation image (Horstmeyer et al. 2009; Su et al. 2015). Such light-field multi-spectral imaging has already been applied alongside with MRT to achieve better temperature measurement accuracy levels (Kelly et al. 2021; Luan et al. 2021).

As schematically shown in Fig. 7.1, a filter array is positioned in front of the main lens (i.e. 3 $\times$ 3 filter as an example shown in Fig. 7.1), such that radiation from a high temperature object is discretised (i.e. 3 $\times$ 3 wavelengths). By optically conjugating the filter and microlens image plane, multi-spectral images of the high temperature object can be captured by a light-field camera with just one snapshot.

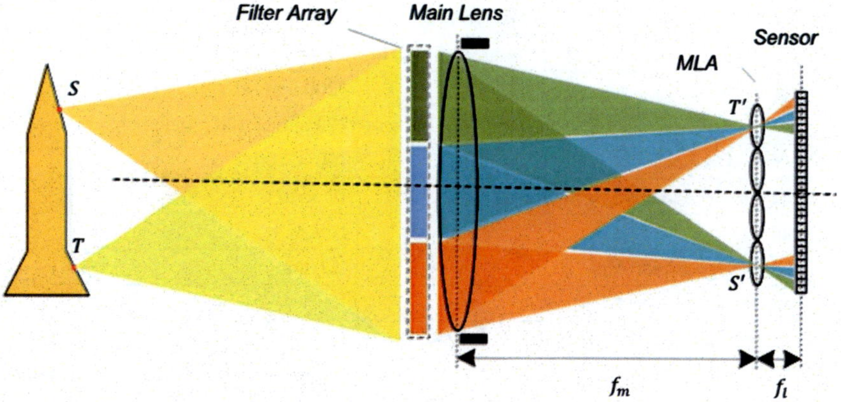

Fig. 7.1 Schematic of a typical light-field-based MRT method. Adapted with permission from Luan et al. (2021) © The Optical Society

To verify this light-field multi-spectral radiation thermometry approach (LF-MRT), synthetic multi-spectral light-field images were generated with the similar ray-tracing method as detailed in Chap. 3. To be specific, synthetic radiational information from an object with temperatures ranging from 500 to 2000 K were propagated to a 3 × 3 filter array during the verification stage. Based on nine selected wavelengths, portions of radiation rays would pass through the main lens and reach the sensor plane. Different from the earlier simulations in Chap. 3, radiation intensity of each ray is regulated according to its wavelength, emissivity of the object and surface temperature, as expressed by Eq. (7.1) (Xing et al. 2017):

$$I_i = A_i \varepsilon(\lambda_i, T) \frac{1}{\lambda_i^5 \left(e^{\frac{C_2}{\lambda_i T}} - 1 \right)} \quad (i = 1, 2, \ldots, 9) \tag{7.1}$$

where A_i is a calibration factor which depends on the wavelength λ_i, camera exposure time, pixel sensitivity, absorption ratio of optical windows and others, $\varepsilon(\lambda_i, T)$ is the spectral emissivity of the target at true temperature T, λ_i is the effective wavelength of the i-th channel and C_2 is the second radiation constant. The final intensity of each pixel was determined based on the total number of rays collected by the pixel and normalised into a range of 0–255. Note that 20 dB white Gaussian noise was added into the raw light-field images for a more realistic approach.

Details of the parameters used to generate synthetic multi-spectral light-field images are listed in Table 7.1, where simulated temperatures ranged from 500 to 2000 K. Additionally, the emissivity values of the nine selected wavelengths were taken from Sun et al. (2005). To better illustrate, a synthetic multi-spectral light-field image generated at $T = 1800$ K is shown in Fig. 7.2a. Each microlens samples 3 × 3 wavelengths of the object radiation, which is then discretised by 12 × 12 image sensor pixels below it, as shown in Fig. 7.2c. Since it is likely that marginal pixels

Table 7.1 Parameters for generating synthetic multi-spectral light-field images

Focus length of main lens	200 mm		
Object distance	2200 mm		
Field-of-view	120 mm × 120 mm		
Microlens number	100 × 100		
Pixel number under single microlens	12 × 12		
Pixel size	10 μm		
Focus length of microlens	1.2 mm		
f-number	10		
Filter array channel (wavelength, λ_i, in μm)	0.574	0.592	0.623
	0.654	0.698	0.748
	0.826	0.870	0.914
Emissivity, $\varepsilon(\lambda_i, T)$	0.409	0.649	0.520
	0.506	0.425	0.416
	0.375	0.371	0.369

Adapted with permission from Luan et al. (2021) © The Optical Society

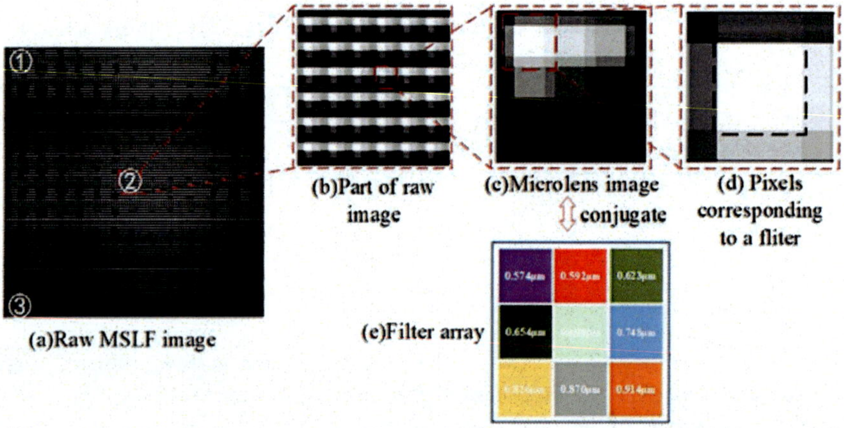

Fig. 7.2 **a** Raw multi-spectral light-field image, **b** part of the raw light-field image, **c** microlens image, (**d**) pixels corresponding to a filter and (**e**) filter array. Adapted with permission from Luan et al. (2021) © The Optical Society

may capture radiation intensities from more than one wavelength due to vignetting effects, and adjacent microlenses may sample radiation from the same wavelength due to "cross-talk", only the centre 2 × 2 pixels in the black dashed-line box shown in Fig. 7.2d were used in the subsequent temperature calculation procedures.

To decouple a raw multi-spectral light-field image, or to recover 3 × 3 spectral channels from one light-field image, procedures similar to multi-perspective image calculation as described in Chap. 5 are adopted. That is, the microlens centre is determined from a board image which is captured without filter array, and pixels beneath each microlens are selected and stitched together according to their relative location to the microlens centre. In this way, 3 × 3 radiation images with a resolution of 100 × 100 pixels are recovered, and nine equations could be established according to Eq. (7.1). Without knowing the emissivity, $\varepsilon(\lambda_i, T)$, the temperature calculation can be converted into a constrained optimisation problem, where the difference between the true and estimated temperatures is to be minimised in accordance with Eq. (7.2):

$$\sum_{i=1}^{n}[T_i - E(T_i)]^2 = 0 \tag{7.2}$$

Based on optimisation methods such as Gradient Projection (GP) algorithm, 100 × 100 data points can be obtained for each temperature. The calculated results are then compared with the true temperature and the relative errors are presented in Fig. 7.3. The measurement error varies between 0.31 and 2.22% and increases with temperature. Due to limitations imposed by the GP algorithm, measurements for higher temperatures will have lower measurement accuracy levels.

The above LF-MRT method may now be compared with traditional RGB-CCD method (Fu et al. 2006) with synthetic multi-spectral images which were generated from the temperature data of a turbine blade. 1870 temperature measurements ranging from 1400 to 1800 K were taken from the blade suction side using thermochromic liquid crystal technique, and the results are presented in Fig. 7.4. Based on these measurement data, synthetic LF-MRT and RGB-CCD radiation images were generated for a 120 mm × 120 mm blade section, as indicated by the black dashed-line

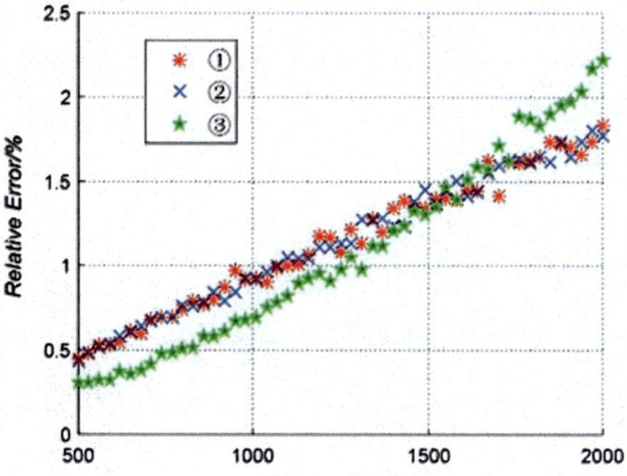

Fig. 7.3 The relative errors between the calculated and true temperatures in areas marked by ①②③ in Fig. 7.2. Adapted with permission from Luan et al. (2021) © The Optical Society

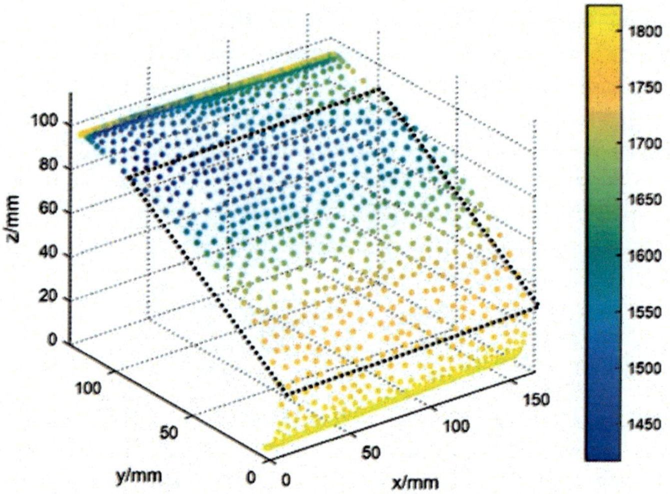

Fig. 7.4 Surface temperature measurement data points on the suction side of a turbine blade. Adapted with permission from Luan et al. (2021) © The Optical Society

box in the figure. 20 dB white Gaussian noise has been added to the synthetic LF-MRT and RGB-CCD images as well, and the final images are shown in Fig. 7.5a, d respectively. After their corresponding post-processing, the evaluated temperature distributions for LF-MRT and RGB-CCD approaches are shown in Fig. 7.5b, e

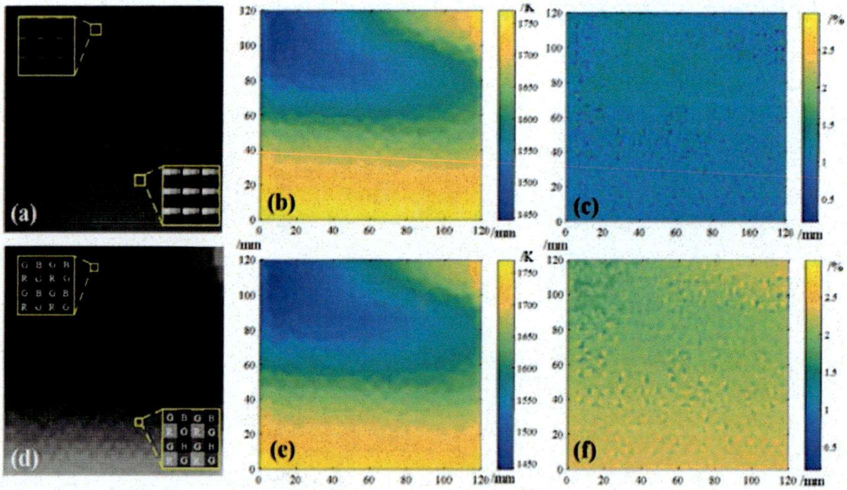

Fig. 7.5 **a, d** Synthetic multi-spectral light-field and RGB-CCD images of a turbine blade suction side (parts of the raw image are shown in yellow boxes); **b, e** evaluated temperature distributions based on LF-MRT and RGB-CCD approaches; **c, f** relative errors of LF-MRT and RGB-CCD approaches. Adapted with permission from Luan et al. (2021) © The Optical Society

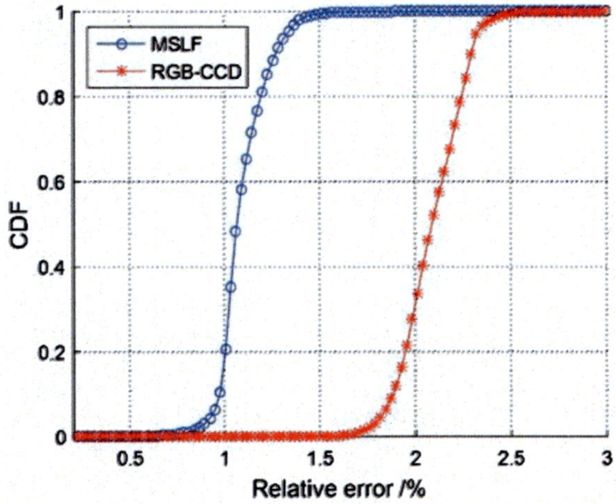

Fig. 7.6 A comparison between the cumulative distribution functions (CDF) of the relative errors associated with LF-MRT and RGB-CCD approaches. Adapted with permission from Luan et al. (2021) © The Optical Society

respectively, which were then subsequently used to determine the relative measurement error distributions. These relative measurement error distributions are shown in Fig. 7.5c, f for LF-MRT and RGB-CCD approaches respectively and demonstrated that, at least for the present turbine blade, that LF-MRT method has higher measurement accuracy levels for the high temperature regions with measurement errors of about 50% lower than the RGB-CCD method. The overall higher accuracy levels of LF-MRT approach can also be appreciated in Fig. 7.6 as well, where the cumulative distribution functions (CDF) of the relative errors associated with both LF-MRT and RGB-CCD are compared directly.

Further experiments were conducted with a black body furnace as shown in Fig. 7.7, where a 9×9 filter array was mounted in front of the main lens of an industrial light-field camera (VOMMA VA4300-M-TGE, 7915×5436 pixels) for calibration purposes. Radiation spectral light-field images were captured for temperatures ranging from 1000 to 1500 K at 470 nm, 500 nm, 530 nm, 575 nm, 620 nm and 670 nm wavelengths. By comparing the calculated temperature with the black body temperature, the measurement error level is estimated to be less than 1%. Studies and experimental testing conducted so far bode well for the widening use of LF-MRT approach in temperature measurement applications that demand contactless, global high temperature distributions at high accuracy levels, particularly applications that offer limited access and yet require ease in the measurement process.

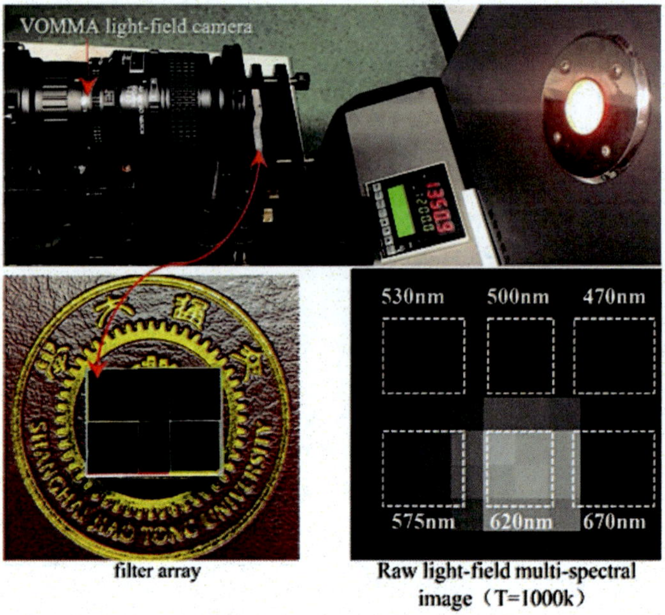

Fig. 7.7 Black body furnace calibration setup with a VOMMA VA4300-M-TGE light-field camera

3D Geometry Measurements

Since light-field cameras have been demonstrated to be capable of reconstructing 3D images for nanoscale and micro-scale PIV tracer particles, it is possible that the 3D geometrical details of solid boundaries and any accompanying volumetric velocity fields could be simultaneously measured with just a single light-field camera. Such 3D fluid–structure-interaction (FSI) investigations were previously accomplished by more conventional multi-view geometry (Adhikari and Longmire 2012; Jeon and Sung 2012; Im et al. 2015) or camera-array-based light-field imaging (Mendelson and Techet 2015). Towards this end, a crucial step other than light-field particle 3D reconstructions, is how to accurately retrieve 3D geometrical details of a model from a raw light-field image. It is worthwhile to highlight also that 3D geometry measurements using a light-field camera is wide-ranging that deserves more attention in the future by itself. However, within the context of this chapter, its coverage here will be more moderate for the sake of brevity.

Although stereo-vision, laser triangulation and structured light techniques are highly efficient and mature for 3D metrology, they may encounter challenges when measuring very small or even tiny 3D physical structures. In particular, limited available viewing angles could result in serious occlusion effects for these techniques. In addition, sharp edges of such small/tiny 3D physical structures are more than likely to drastically affect laser line reflectance or image patterns that are been projected from any laser emitters or structured light projectors. A good example of a 3D metrology scenario that exhibits such challenges will be that associated with turbine blade film cooling holes. To meet the ever-increasing desire to achieve high thrust-to-weight ratios, turbine blades have to work safely under extremely high gas temperatures. Take for instance, the M88-2E4 jet engine that powers the French multi-role fighter RAFALE has a turbine inlet temperature of about 1900 K (Desclaux and Serre 2003). It is predicted that the turbine inlet temperature of next generation jet engines could reach as high as 2300 K. To protect turbine blades and vanes from burning out by such high temperatures, coolant air from the compressor stage is fed into the root of these turbine blades. The coolant air will flow through the internal cooling channels as shown in Fig. 7.8 and ejected out from tiny holes distributed along the turbine blade surface. This produces a classical "jets-in crossflow" like flow behaviour that will see a thin layer of coolant air being formed along the turbine blade surfaces and shield the latter from the much hotter flow over the blade surfaces.

It should be emphasised that geometrical details of these tiny film-cooling holes such as their ejection angles, orientations, shapes and exit areas are specially designed according to stringent aerodynamic and thermodynamic requirements, so as to ensure the highest possible cooling efficiency (Gritsch et al. 2001, 2005; Bohn and Krewinkel 2009; Montomoli et al. 2012). Usually, film-cooling holes are nominally very small with diameters ranging from 0.2 to 0.8 mm, and relatively deep with depth-to-diameter ratios ranging from 2 to 6. These physical attributes make the probes of

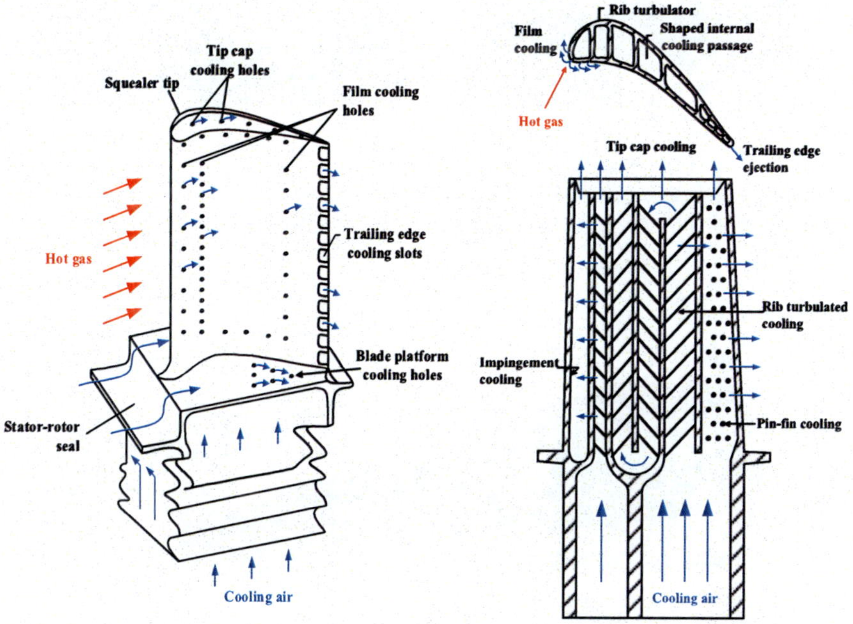

Fig. 7.8 Schematics of how coolant air enters the turbine blade via the blade root, flows through the internal channels and exhausts out of the blade surfaces for film-cooling applications. Adapted from Han and Rallabandi (2010) CC BY 3.0

coordinate-measuring machines (CMM) too large to be used meaningfully. At the same time, the very narrow available viewing angles of these small but deep film-cooling holes would result in significant occlusion effects for stereo-vision, laser triangulation and structured light approaches. Early studies showed that light-field cameras may provide a feasible alternative for such geometrical measurements, since it can achieve 3D measurements despite limited viewing angles (Ding et al. 2019; Zhao et al. 2020).

An example of a light-field camera-based measurement system is shown in Fig. 7.9a, which consists of an industrial light-field camera (VOMMA VA4300-M-CL, 7915 × 5436 pixels) and a main lens (i.e. Micro-Nikkor 200 mm lens coupled together with a Nikkor 50 mm lens). After acquiring the raw light-field image of the turbine blade as shown in Fig. 7.9b, the next step will be to extract its multi-perspective images. As illustrated schematically in Fig. 7.10, rays from different viewing angles are represented by different colours. For the sake of simplicity, only

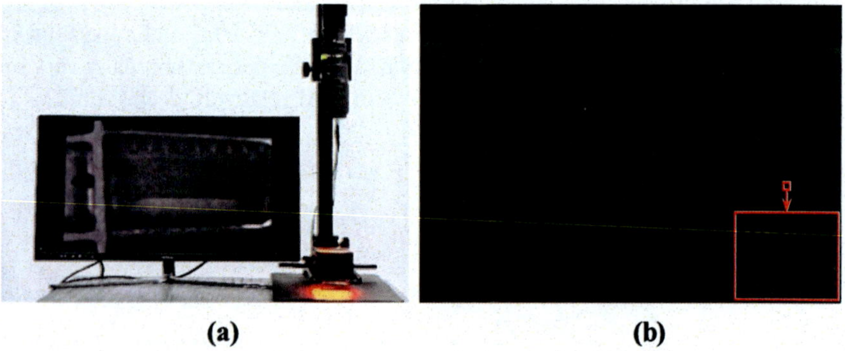

Fig. 7.9 **a** Light-field camera-based turbine blade and film-cooling hole measurement setup, and a **b** raw light-field image of the turbine blade surface (inset shows a magnified view)

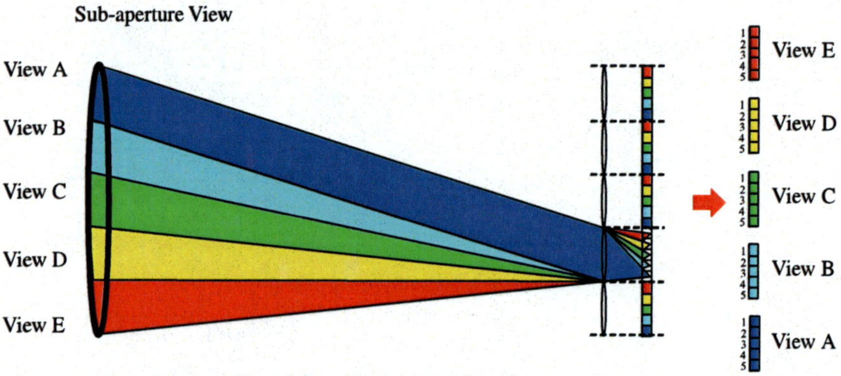

Fig. 7.10 Schematic of multi-perspective image generation from a raw light-field image

5 pixels are shown in this one-dimensional (1D) schematic (whereas it will be 5 × 5 pixels for a two-dimensional (2D) case), which means that a total of 25 perspective images can be generated based on light-field imaging principles. As an example, if one takes the 5th pixel beneath each microlens (i.e. the blue pixels) and stitches them together, a top view could be generated (i.e. View A).

With the availability of these 25 perspective images, a 5 × 5 image matrix can be formed as shown in Fig. 7.11a. If one row of pixels from each of the 25 perspective images are stacked together, an epipolar-plane-image (EPI) can then be generated as shown in Fig. 7.11b. A prerequisite of accurate 3D geometry measurements is how to correctly estimate depth information from an EPI. Fundamentally speaking, slopes of various lines in an EPI correspond to the disparity between different perspective images. Accurate prediction of EPI line orientations is therefore the key task for light-field-based 3D measurements, which could be accomplished by traditional EPI

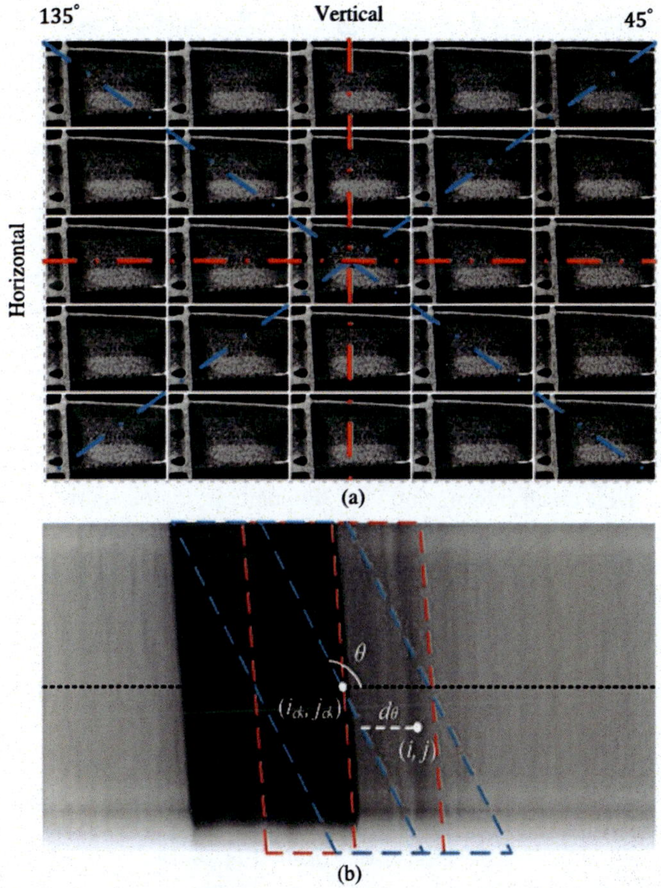

Fig. 7.11 **a** Multi-direction EPI method and **b** compass operator-based slope detection method

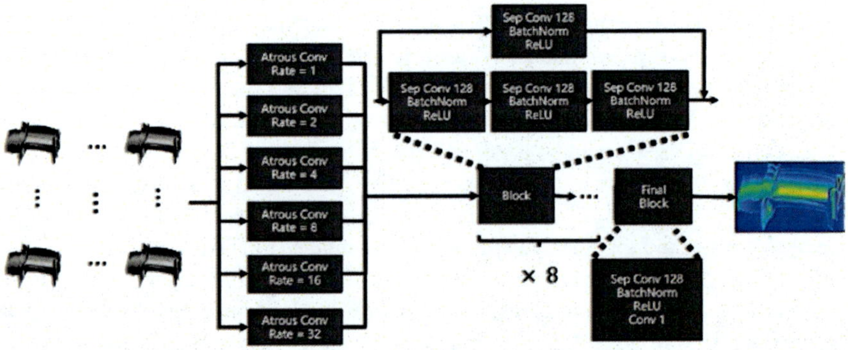

Fig. 7.12 Structure and workflow of end-to-end deep learning network for turbine blade depth recovery. Adapted from Ma et al. (2019) CC BY 4.0

algorithms (Johannsen et al. 2016; Sheng et al. 2018) and EPI-based neural networks (Heber et al. 2017; Feng et al. 2018; Shin et al. 2018). However, multi-perspective images that are generated from raw light-field images taken of turbine blades or film-cooling holes have reflective and texture-less areas, which would greatly affect the depth recovery accuracy levels. To overcome such difficulties, an end-to-end deep learning network could be applied (Ma et al. 2019).

As shown in Fig. 7.12, a deep learning network could use astrous convolution with different dilation rates to extract multi-scale features from all multi-perspective images, so that disparities of reflective and texture-less areas can be estimated from their neighbouring regions. Low level multi-scale features from such feature pyramid are then processed by a series of residual blocks, each comprising of a shortcut depth-wise separable convolution layer and three consecutive depth-wise separable convolution layers. Lastly, accurate depth information of the turbine blade surface or film-cooling hole is calculated from the final residual block.

The volumetric calibration algorithm introduced in Chap. 5 may also be applied here to convert depth information into 3D points in metric units. Performance of such a procedure can be accessed with standard gauge blocks. Table 7.2 shows the measurement accuracy levels at different depths-of-field and fields-of-view for a process capability index of $C_{pk} \geq 1.33$. Applying metric calibration to the turbine blade depth image captured by a light-field camera as shown in Fig. 7.13a, a very dense 3D point cloud can be obtained, as can be seen in Fig. 7.13b with surface texture included. To be more precise, approximately 820,000 3D points were obtained

Table 7.2 Measurement accuracy of a VA4300-M-CL light-field camera system

Magnification	1.5×	2×	4×
Depth-of-field (mm)	5.1	3.1	0.78
Field-of-view (mm)	14.8 × 10.1	11.1 × 7.6	5.5 × 3.8
z-accuracy (μm)	±37	±17	±8
xy-accuracy (μm)	±3	±3	±1

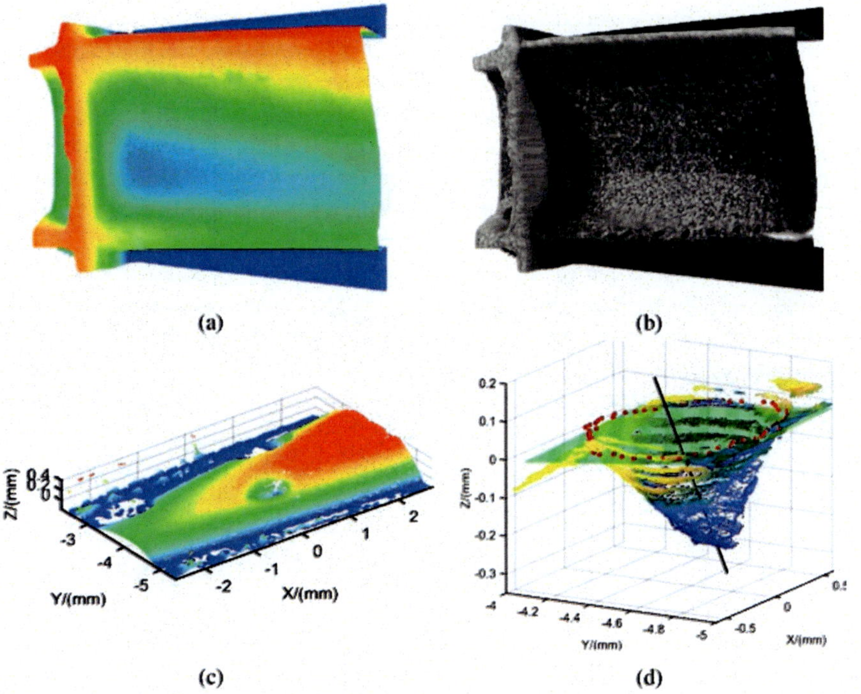

Fig. 7.13 **a** Depth image of turbine blade, **b** 3D point cloud of turbine blade, **c** Depth image of a film cooling hole, **d** 3D point cloud of a film cooling hole

through the use of a VA4300 industrial light-field camera. If one were to zoom in on a single film-cooling hole with a 4× lens, a very dense and highly accurate point cloud can be obtained, as shown in Fig. 7.13c. Based on these data points, Fig. 7.13d shows how details of key geometries such as ejection angle, orientation, shape and exit area of the film-cooling hole may be realised.

Droplet and Splash Measurements

One of the first pseudo-3D flow imagings and hence structural reconstruction arose from stereoscopic imaging was conducted by Ng and Zhang (2003), where they made use of a stereoscopic adapter mounted in front of the lens on a 2D digital camera. The two sets of mirrors within the adapter allow the imaging of the same scene via two different perspectives by the CCD sensor, which in turn produce two different views in a single image. As the geometrical and optical details are known quantities, the two different views can be combined to produce a 3D reconstructed view of the scene. Similar principles had also been demonstrated by Wang and Zhang (2011) and Wu et al. (2020) on droplet/splash research, though in this case they reconstructed 2D shadowgraph image pairs derived from a stereoscopic shadowgraph system instead. To illustrate briefly, both studies made use of a stereoscopic shadowgraph system shown in Fig. 7.14, where the scene of interest is located within the shaded test section region. The relationship between a point within the scene and its images on the same sensor in the case of a stereoscopic adapter based setup, or two separate sensors arranged stereoscopically, may then be represented by Fig. 7.15. This relationship can then be described mathematically for reconstruction of the 3D images and used to track regions or areas of interest.

A good example can be seen in a bubble undergoing bursting process imaged stereoscopically by Wang and Zhang (2011). The images were taken based on a shadowgraph system comprising of a 500 W xenon light source and four ~300 mm parabolic mirrors with ~3 m focal lengths, where two Photron SA-3 monochrome high-speed cameras were used to capture the shadowgraph images simultaneously from two different perspectives. To reconstruct the droplet trajectories after the bubble

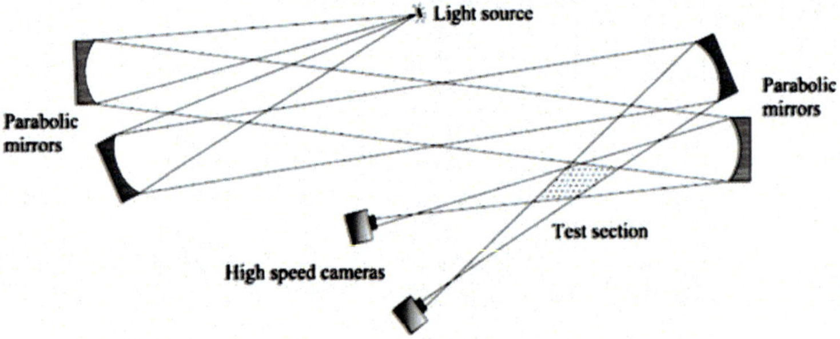

Fig. 7.14 Schematics of a typical stereoscopic shadowgraph setup (Wang and Zhang 2011)

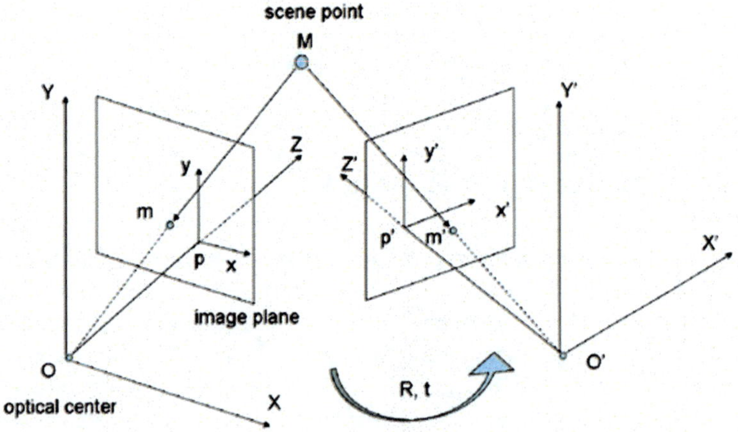

Fig. 7.15 Relationship between the physical coordinates and image coordinates of the two images resulting from stereoscopic imaging (Wang and Zhang 2011)

had burst, calibration of both cameras would have to be carried out, such that the depth information can be quantified from the images captured by the two different arranged cameras. This was done by having both cameras to take a photo of a planar calibration board with a known pattern and dimensions each as shown in Fig. 7.16, such that the relationships between the physical and images coordinates could be established for stereoscopic reconstructions (Zhang 2000). For more details on the rationale and formulation of the stereoscopic reconstructions, readers are referred to Wang and Zhang (2011).

The cameras were set to 5000 FPS for the experiments and triggered simultaneously, where a bubble of approximately 44 mm was punctured. The captured image sequences for both cameras are shown in Fig. 7.17, where they are separated by a

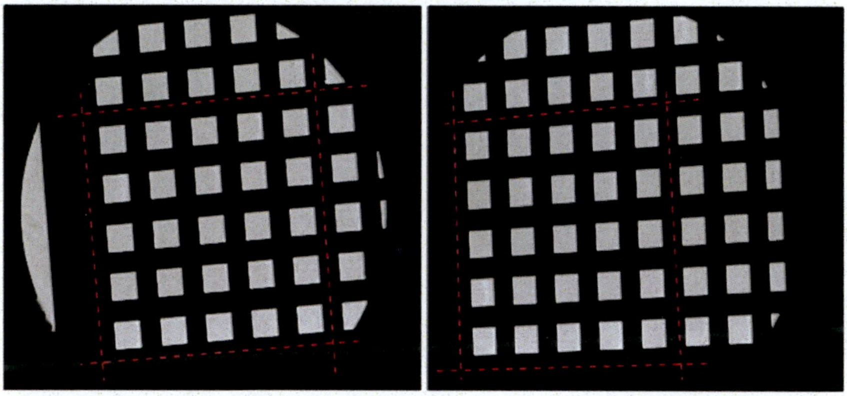

Fig. 7.16 Image samples of the calibration board taken by stereoscopically-arranged cameras, where each sample was taken from a different viewing perspective (Wang and Zhang 2011)

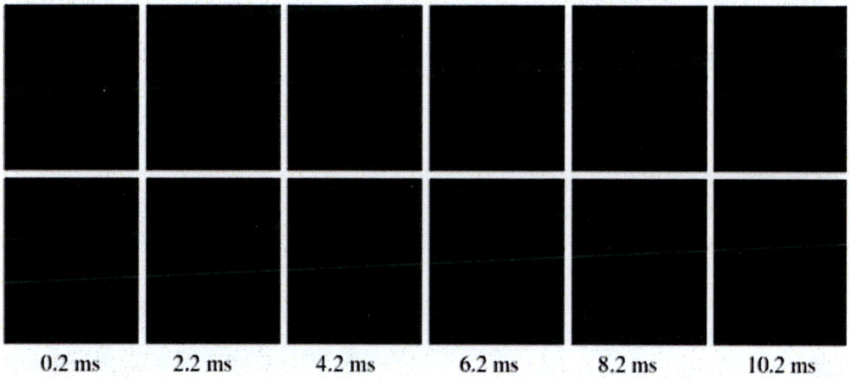

Fig. 7.17 Stereoscopic images captured of a bursting bubble, where the top and bottom rows show the images captured from two different perspectives (Wang and Zhang 2011)

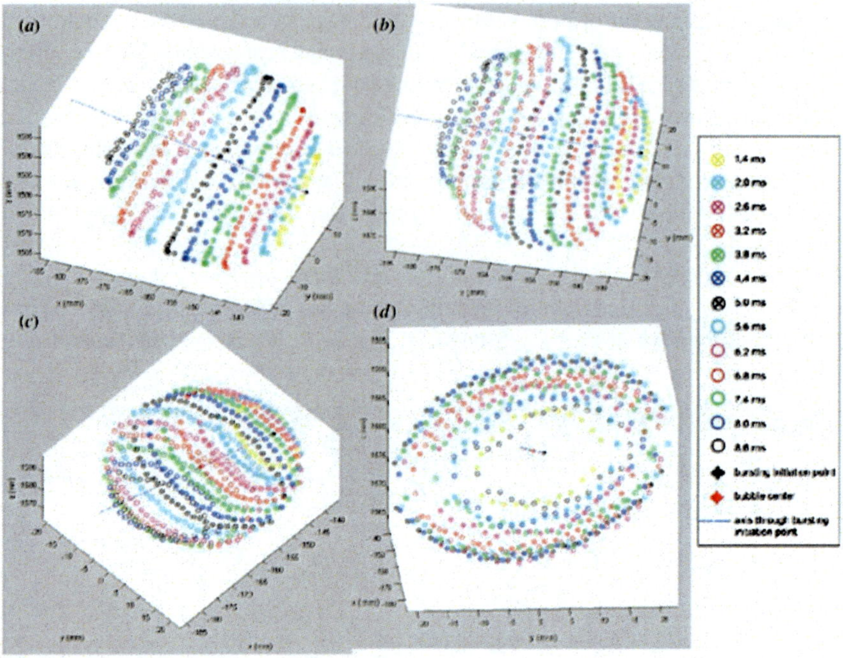

Fig. 7.18 Temporal evolution of the bubble bursting boundary presented in 3D views after stereoscopic reconstructions (Wang and Zhang 2011)

time-interval of 2 ms. Whilst the two sets of image sequences do provide a qualitative grasp of the bursting phenomenon, they do not directly provide data that can be analysed quantitatively. However, if the calibration data was available such as in the case for (Wang and Zhang 2011), the boundaries of the bubble rim during the bursting process could be quantitatively extracted and plotted as shown in Fig. 7.18. It can be seen that the availability of these 3D results made it possible to track the evolution of the bubble bursting process as a function of time in a more systematic and insightful manner. However, as noted in that paper, it is occasionally necessary to select some of the points for stereoscopic reconstruction manually, due to the occasional arbitrariness in the selection of a matched feature in both images. This is due to the fact that 2D cameras without depth information were used, and user interventions based on human depth perception would have to come in.

Nevertheless, such a stereoscopic setup was very useful not only in terms of tracking a continuous bubble rim during the bursting process, it was also found to be satisfactory when it comes to tracking the droplets formed at the same time. For instance, Fig. 7.19 shows that the trajectories of selected droplets produced during the bursting of a bubble in 3D space, which show the feasibility of the stereoscopic reconstruction approach.

Wu et al. (2020) improved upon the preceding stereoscopic reconstruction approach when they applied a relatively similar shadowgraph imaging setup for their

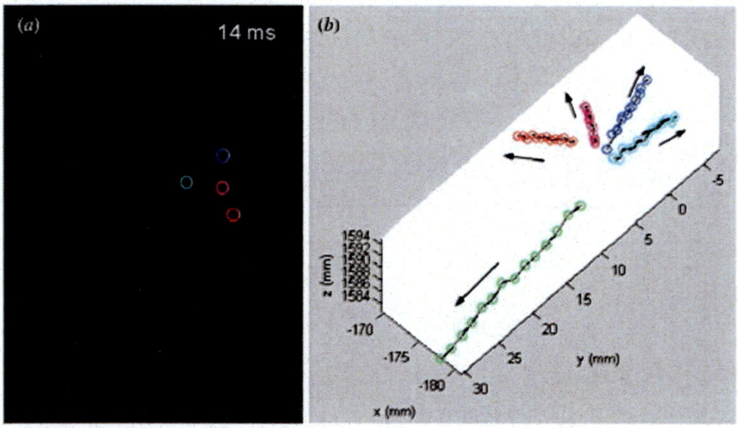

Fig. 7.19 3D trajectories of selected droplets produced during the bursting of a bubble over a period of 4.8 ms (Wang and Zhang, 2011)

splash-induced droplets study. Several major improvements were made to the droplet tracking and stereo-matching however—Firstly, the tracking of droplets between consecutive image frames was enhanced by the implementation of a 3D particle tracking velocimetry algorithm to more reliably determined the droplet 2D positions. These were done separately for the sequential images captured by each of the two high-speed cameras. Secondly, epipolar constraint was used in the stereo-matching of the droplets from images captured by the two cameras. Mathematically speaking, a droplet in an image from one camera should lie on the epipolar line of the corresponding image from the other camera. This essentially meant that the search for the droplets in corresponding image frames was restricted to certain regions of the images (with some small tolerances built in), thereby reducing the likelihood of stereo-matching the wrong droplets between the two images. Note that only 2D droplet trajectories that were able to fulfill the epipolar restrictions would be used for subsequent stereoscopic reconstructions. Last but not least, stereoscopic reconstructions were carried out based on the calibration and reconstruction procedures described earlier on. However, additional steps were incorporated such that droplet trajectories which did not seem reasonable would be rejected. For the present scenario where droplet density was relatively low, this remains feasible. However, it remains unclear if it will continue to be feasible if droplet density increases significantly. An illustration showing the above procedures is presented in Fig. 7.20.

To show what the approaches adopted by Wu et al. (2020) were able to accomplish, Fig. 7.21 first shows how the splash droplets were formed by 2D shadowgraph visualisations under a variety of Weber number and film thickness conditions. It is clear from these results that the different crown-type and crown bottom splashing produce a wide range of droplet formations and trajectories, and Fig. 7.22 shows the 3D trajectories of the secondary droplets made possible by the procedures laid out in Fig. 7.21. It can be seen that the present procedures were able not only to

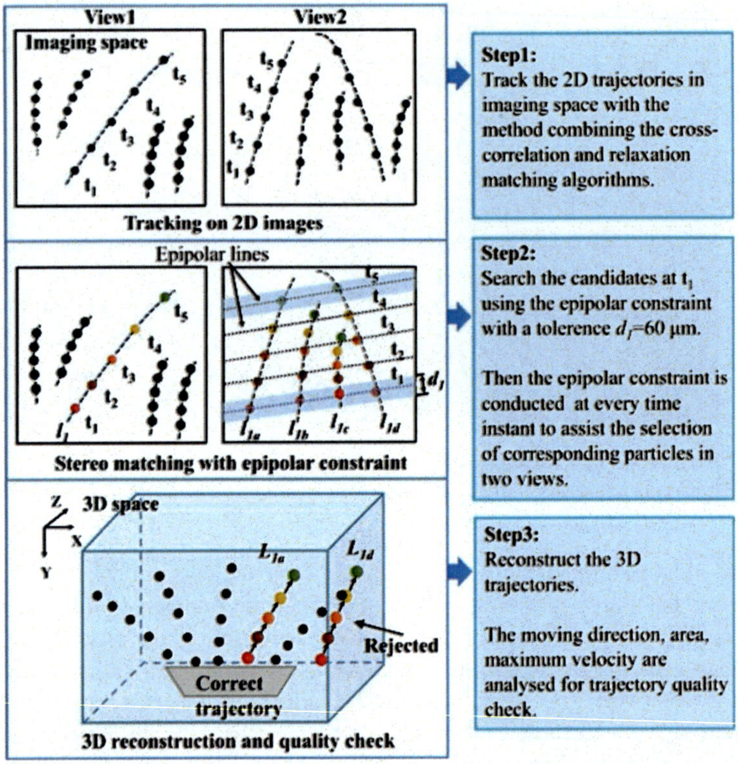

Fig. 7.20 Droplet tracking, stereo-matching and reconstruction schemes that were used to track the droplet 3D trajectories (Wu et al. 2020)

differentiate a significant number of distinct droplets, but to quantify their velocity magnitudes as well. From a technical standpoint, the results are highly convincing and reveal much on the mechanisms of droplet formation and behaviour associated with the splashing phenomena investigated during the study.

On the other hand, the need to have two high-speed cameras synchronised to acquire image sequences from two different viewing perspectives may prove to be a challenge if two similar cameras (i.e. usually recommended) are unavailable, or if cost is a primary concern. Additionally, the need to carry out stereo-matching to correctly identify the same droplet across two different images before tracking its 3D trajectory may also present certain experimental uncertainties. Note that the tolerance set for the droplet search area could also be subjected to human interpretations. Hence, utilising a single light-field camera may alleviate some of the above issues. Take for instance, firstly, it presents itself as a single camera solution where the calibration process is likely to be more straight forward and camera adjustments for different viewing perspectives would be simpler. Secondly, there is no need to carry out the stereo-matching step, since only one camera will be imaging all the droplet formations and trajectories—there is only a single series of light-field image sequence

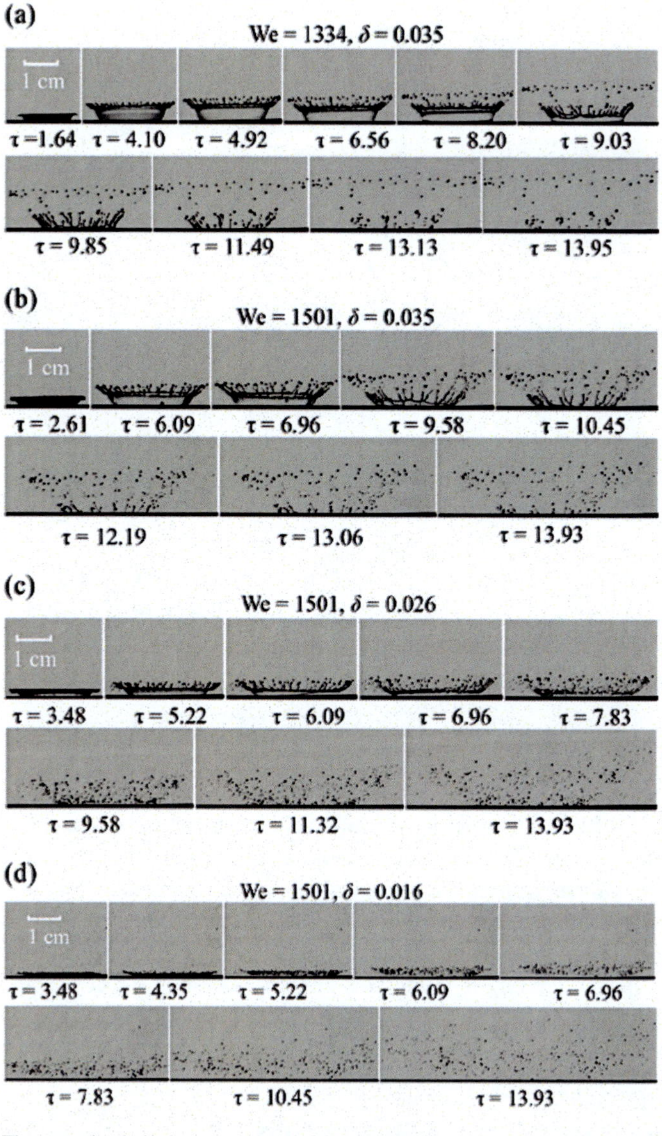

Fig. 7.21 Time-resolved shadowgraph-based visualisation images for splash droplet formations for various Weber numbers and film thicknesses (Wu et al. 2020)

for each experiment and there will be lower uncertainties when it comes to droplet identifications. Thirdly, 3D trajectory tracking can be accomplished by making use of 3D multi-grid cross-correlations similar to what had been adopted for LF-PIV procedures. With these in mind, it would appear that a light-field camera can simplify the required imaging setup and reduce the experimental cost significantly. Having said that, the high framerate requirement for time-resolved droplet investigations

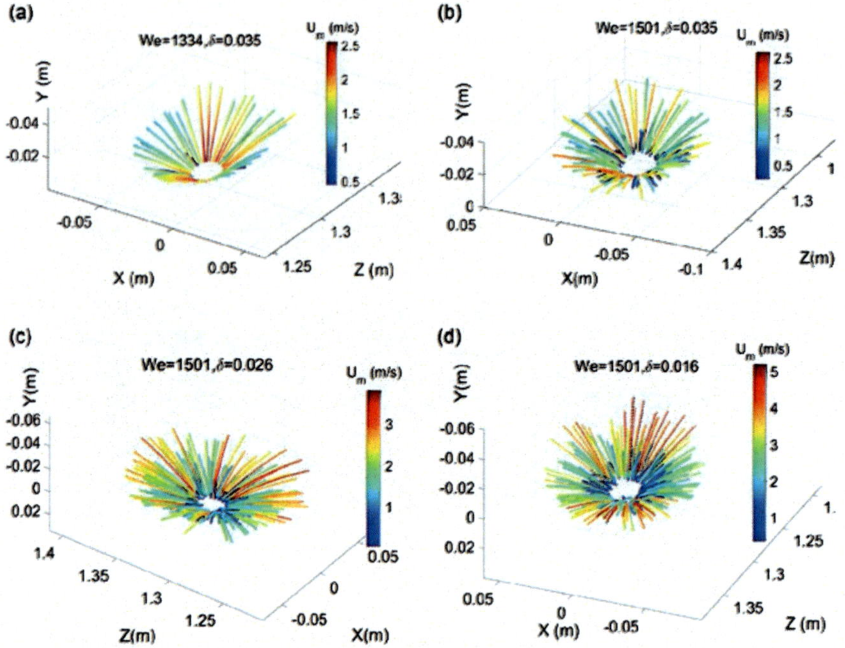

Fig. 7.22 3D trajectories of the secondary droplets corresponding to the cases visualised in Fig. 7.8 (Wu et al. 2020)

will pose technical challenges for a light-field camera-based setup. In particular, the light-field camera will need a very high resolution and hence pixel count imaging sensor for accurate 3D reconstructions of the droplets.

Whilst imaging sensor technological advances meant that resolution continues to increase, current speed limitations in the sensor readouts limit the framerate of a high resolution light-field camera to about 100–200 FPS, instead of an (or more) order-of-magnitude higher framerates offered by current lower resolution 2D high speed cameras.

3D Flame Structure Reconstructions

Combustion phenomena and associated behaviour under a variety of fuel mixtures, nozzle/holder geometries and other operating conditions have always been an area of intense research, due to the tremendous need to achieve more efficient combustion processes, lower carbon emissions, reduced noise emissions and higher performances, to name a few desirable outcomes. Until recently, a significant number of experimental investigations made use of 2D or stereoscopic PIV, as well as laser-induced fluorescence based and other tools, to capture quantitative results associated with combustion and/or flame behaviour, and some recent studies would include those by Stratton (2005), Filatyev et al. (2007), Petersson et al. (2007), Hartung et al. (2009), Hossain et al. (2013), Wang et al. (2014, 2015, 2020), Weinkauff et al. (2015) and Huang et al. (2020), amongst others. As a result, it is not surprising that strong desires to understand and correlate full 3D flame structures to the eventual combustion efficacy exist. This also raises the possibility of reconstructing flames structures with sufficiently high density-gradients based on shadowgraph or Schlieren imaging much like the previous section.

One of the earliest studies on the stereoscopic reconstruction of 3D flame surface features will be the one conducted by Ng and Zhang (2003). In their study, they made use of a prismatic stereoscopic adapter mounted upon a digital camera that allowed the capture of the same scene from two different viewing perspectives onto one single digital image, as shown in Fig. 7.23. To reconstruct the 3D image with depth information, they made use of relatively similar stereoscopic calibration and reconstruction principles as those adopted by Wang and Zhang (2011). The primary difference between the approaches by Ng and Zhang (2003) and Wang and Zhang (2011) is that the latter made use of two cameras instead of one by the former. Nevertheless, the core principles remained the same. For the former study, they made use of an impinging coaxial-type burner, where propane gas underwent combustion with air. A flat steel plate was used as the impinging surface, and the primary interest was in the capturing of the flame structures produced along it. A digital camera with

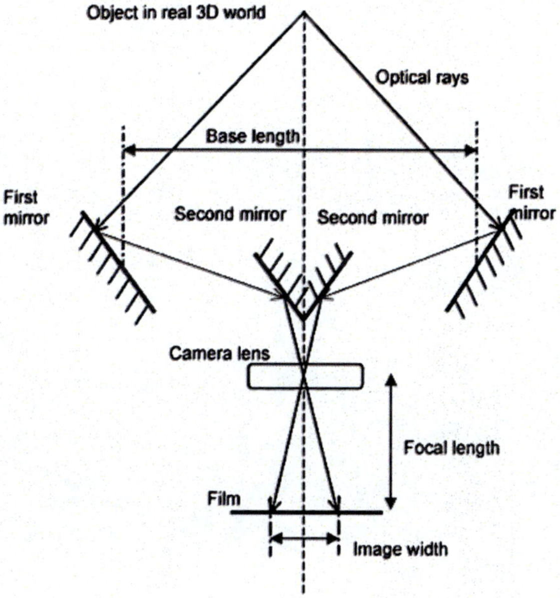

Fig. 7.23 Schematics of the basic operating principles of a stereoscopic adapter used in Ng and Zhang (2003)

the stereoscopic adapter was mounted on a tripod with its optical axis pointing at the flames along the lower steel plate.

The top row of Fig. 7.24 shows the two images of the flame structures captured by the camera stereoscopically, whilst the second and third rows show the reconstructed 3D flame structures from different viewing perspectives. The availability of reconstructed 3D flame structures mean that they could be rotated for a much more meaningful visualisation and interpretation. Note that these results actually mapped out the external flame structure surfaces, rather than the phenomena occurring within the flame structures. Nonetheless, this provides a good opportunity to quantify and map out the flame surfaces, such that the localised flame surface areas and surface densities could be estimated. In this case, depth information from the current experimental approach and Delaunay triangulation-based polyhedral approximations of the surfaces were utilised to map out the flame surfaces continuously. The latter triangulation step was needed to interpolate the 3D surface distributions between the points for which depth information was available, for a more realistic surface representation. Figure 7.25 shows how some of the 3D meshes that were based on flame surfaces of interest looked like. Note that since the calibration process would have provided information on the magnification factor, it would be possible to estimate the physical areas of selected flame surfaces and the availability of flame surface areas could then be used to determine other combustion parameters of interest, such as fuel consumption rate, take for instance.

An improved optical-based approach of 3D flame structure reconstruction for impinging flames based on a relatively similar setup was demonstrated by Wang et al. (2013). In this case, the flame structures were not imaged directly but instead

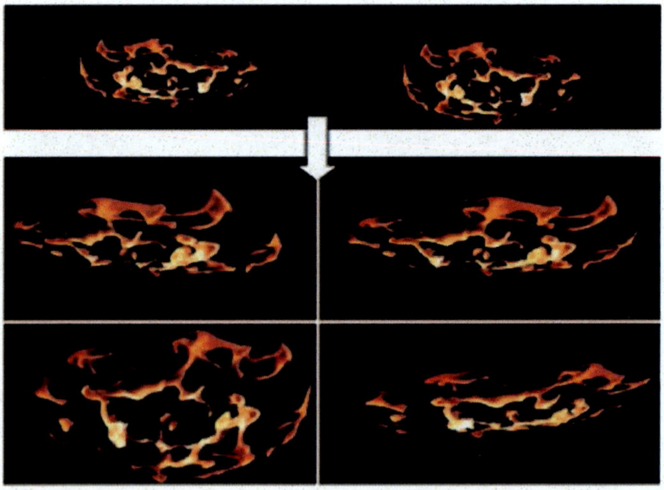

Fig. 7.24 A sample pair of stereoscopic images captured by the camera (top row) and how the reconstructed 3D flame structures look like when viewed from different viewing orientations (second and third rows) (Ng and Zhang 2003)

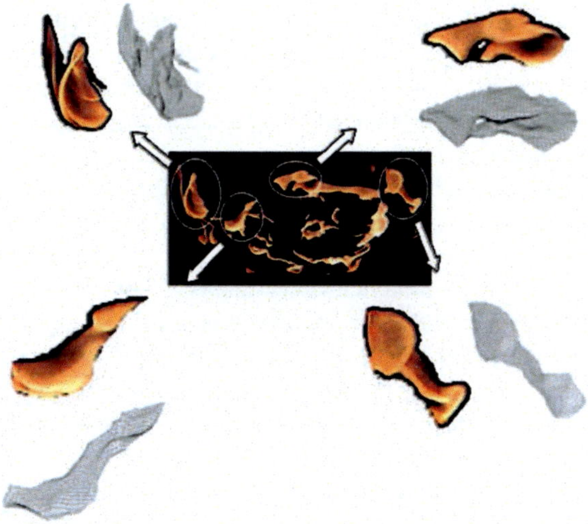

Fig. 7.25 Reconstructed 3D meshes based on selected flame structure surfaces (Ng and Zhang 2003)

through the use of a Schlieren system, such that the density gradients were visualised. Hence, there are some distinct differences in terms of identifying the flame structures between this study and (Ng and Zhang 2003). Figure 7.26 shows the experimental setup used by Wang et al. (2013), where there were two high-speed cameras

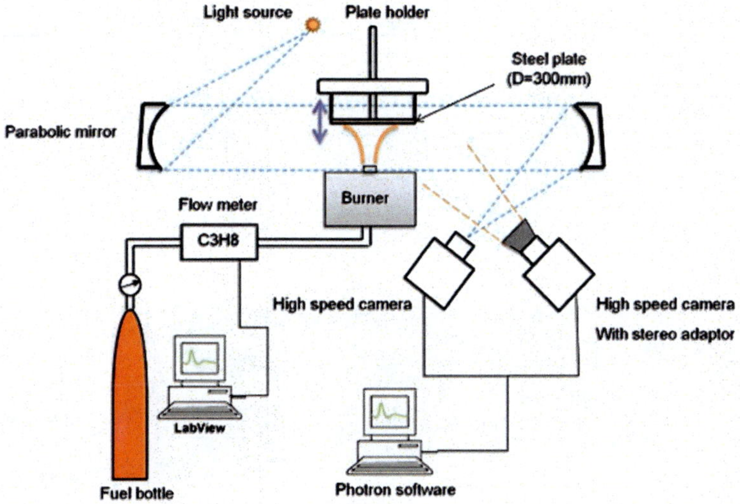

Fig. 7.26 Schematics of the experimental setup used by Wang et al. (2013). Note that one camera was used for visualisation, whilst the other was used for stereoscopic imaging through the use of a stereoscopic adapter

capturing the Schlieren images. Note that one camera was capturing the 2D schliren visualisations qualitatively, whilst the the other camera was fitted with a stereoscopic adapter similar to what (Ng and Zhang 2003) had done. Such a setup offered the advantage of obtaining simultaneous visualisations and measurements, the results of which could be interpreted concurrently for a better correlated understanding. More interestingly, as part of flame feature enhancements, the flames were identified based on a digital colour analysis approach. In particular, colour signals from the captured digital images were extracted and post-processed to separate the soot-induced and radical chemilumininescence-induced flame components. The former and latter components would show up as yellowish-red and greenish-blue regions in the post-processed images, which would aid in diving more deeply into the details of the combustion scenario at head, such as how chemilumininescence varies with respect to equivalence ratio and heat release rate, take for example.

Of particular interest to the study would be the flame structures corresponding to chemilumininescence colour signals such as yellow and blue colours. This was implemented on the images captured for three different ignition positions, where positions 1, 2 and 3 were located at the burner exit, midway between the burner exit and impingment plate and adjacent to the impingement plate lower surface, and Fig. 7.27 shows the time-sequenced blue-colour enhanced images captured for these three positional configurations. It can be clearly discerned from the figure that the

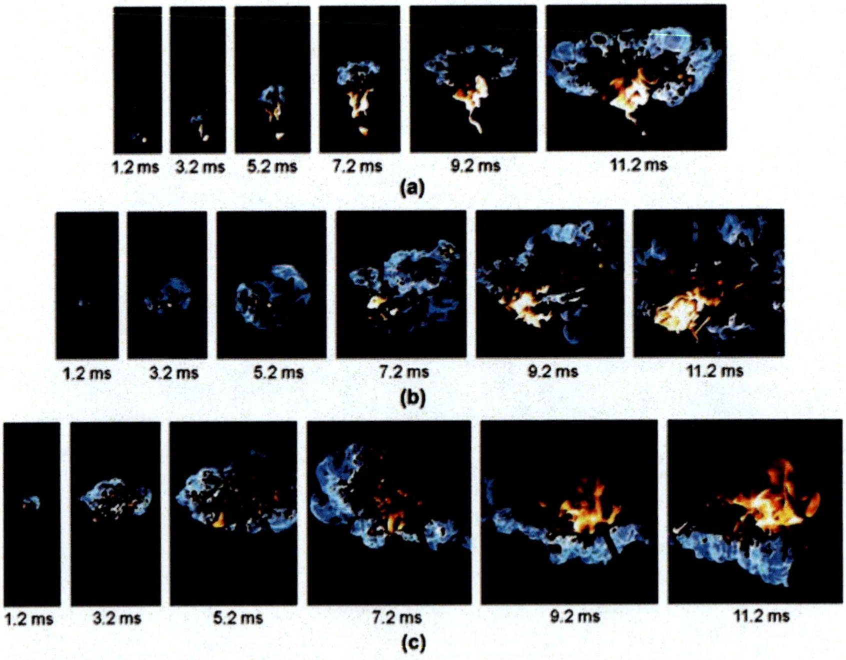

Fig. 7.27 Time-sequenced blue-colour enhanced images captured for **a** position 1, **b** position 2 and **c** position 3 ignition configurations (Wang et al. 2013)

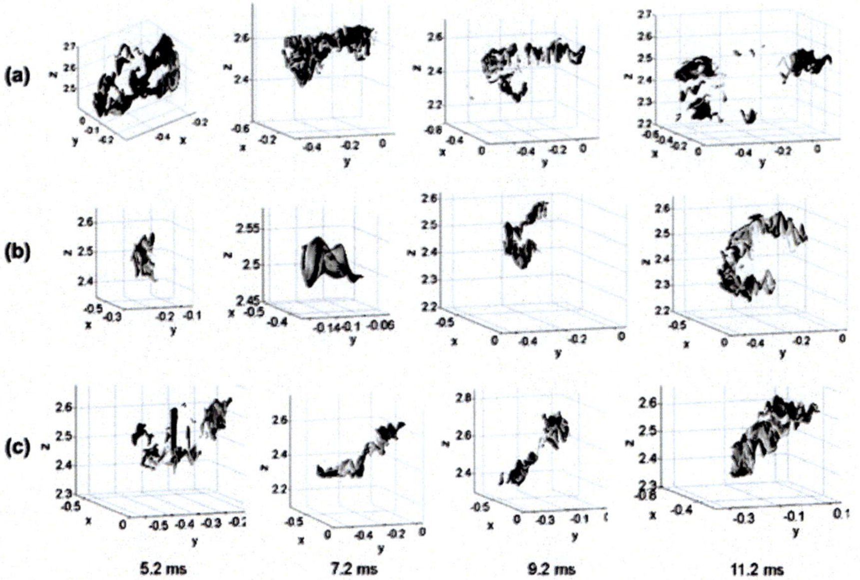

Fig. 7.28 Reconstructed 3D flame structures based on blue-colour enhanced stereoscpically-captured images for **a** position 1, **b** position 2 and **c** position 3 ignition configurations (Wang et al. 2013)

locations at which the fuel/air mixture was been ignited play important roles in the resulting flame structures and combustion behaviour. Stereoscopic reconstructions were subsequently carried out on the steroscopically captured image sets and the resulting reconstructed flame structures for the three different ignition positions are presented in Fig. 7.28.

Success in the study prompted a follow-up study by Wang et al. (2015) whereby both the blue- and yellow-colour enhanced images were reconstructed. The experimental setup and procedures were similar to Wang et al. (2013) and hence they will not be elaborated here again. The same goes for the reconstruction procedures. What is worthwhile to highlight here is the reconstruction of the yellow-colour enhanced images, where very different flame structures and behaviour exist, as can be observed in Fig. 7.29. Nevertheless, the reconstructions of the 3D flame structures for this case were possible and the results are shown in Fig. 7.30.

A significant improvement to the preceding experimental setup and procedures was reported by Yu et al. (2020), where a larger number of images from different viewing perspectives were used to reconstruct the 3D flame structures. Additionally, a more comprehensive 3D reconstruction approach was implemented for the chemiluminescence signals, as compared to the studies described earlier. In particular, the reconstruction procedures relied on algebraic reconstruction technique (ART) commonly used in computed tomography of chemiluminescence (CTC) and readers are referred to Floyd and Kempf (2011), Moeck et al. (2013) and Yu et al. (2018, 2020) for more details on the reconstruction formulations. The study involved two important

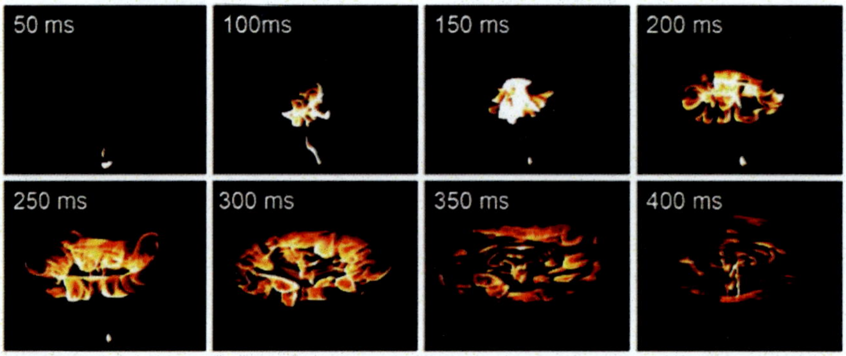

Fig. 7.29 Yellow-colour enhanced images of the methan diffusion impinging flame (Wang et al. 2015)

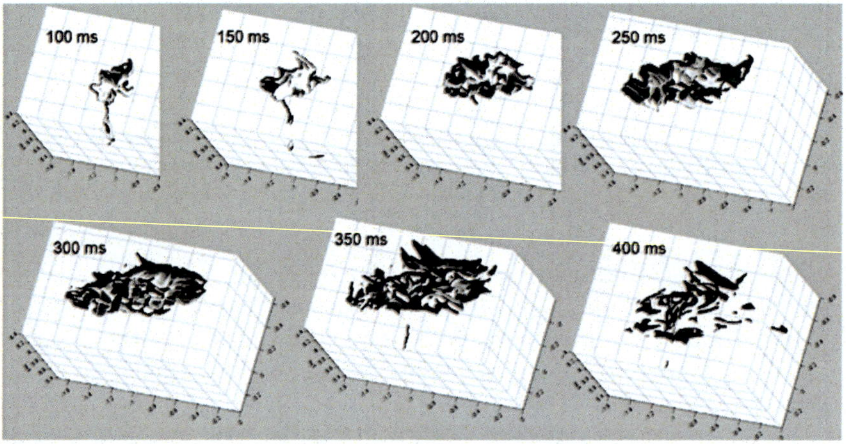

Fig. 7.30 Reconstructed 3D flame structures based on yellow-colour enhanced stereoscpically-captured images corresponding to Fig. 7.29 (Wang et al. 2015)

considerations related to the present scope—firstly, a comparison on how the camera number affects the fidelity of the reconstruction results and hence, the correct identification of flame structures, and secondly, the implementation of a point-extract method rather than threshold method to identify the flame surfaces. Figure 7.31 shows the experimental setup used in this study, where the combustor was designed to produce a turbulent swirling flame. Note that two high-speed cameras were used like in the studies discussed earlier, but in this case, each camera was connected to a four-input, one-output fibre bundle. Hence, all eight fibre bundle inputs coupled with lenses were located at different viewing angles with respect to the flame, where θ ranged approximately from 46° to 131°. Each fibre bundle would then transmit images from four different viewing angles to each camera, where each image would contain four different sub-images. With both cameras operating at a synchronised

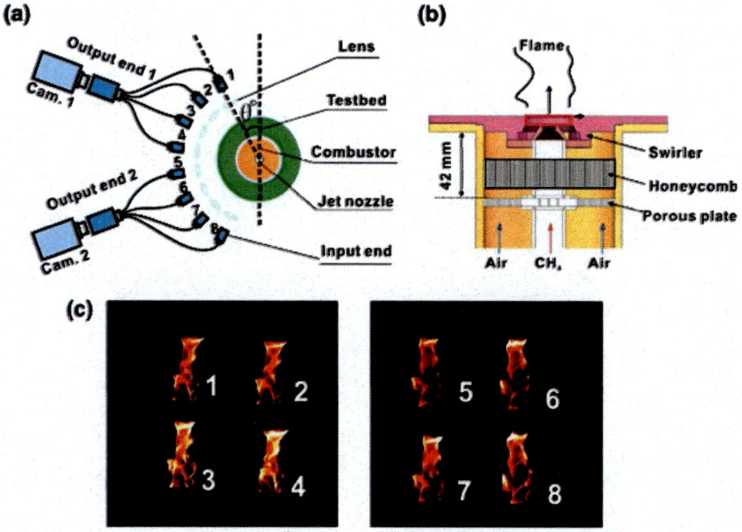

Fig. 7.31 Schematics of the experimental study that show **a** the two high-speed cameras and fibre bundles arrangement, **b** combustor design and **c** image-pair captured simultaneously by the two high-speed cameras (Yu et al. 2020)

4 kHz, these images were subsequently used to reconstruct the transient 3D flame structures.

A comparison between the number of images with unique viewing angles used for the 3D flame reconstruction and the fidelity of the reconstructed flame structures can be seen in Fig. 7.32, where the results for 5–8 images are shown. Whilst the primary flame structure remains invariant across the different camera numbers, some subtle differences can be discerned in the highlighted smaller and peripheral structures. The authors determined the correlations between 5 and 8 images, 6 and 8 images and 7 and 8 images, and found that satisfactory convergence could be achieved with

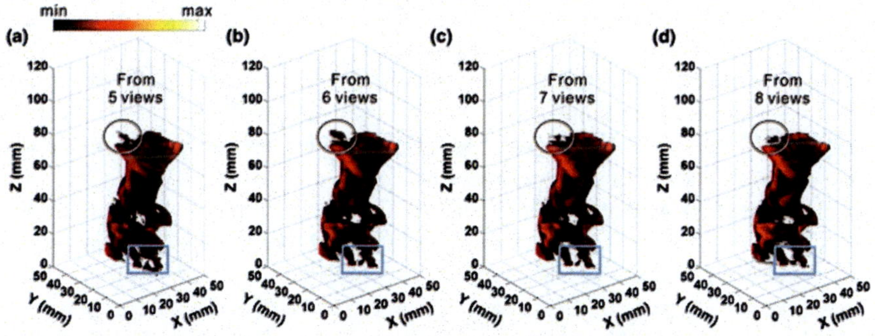

Fig. 7.32 Reconstructed 3D turbulent swirling flame structures resulting from the use of 5–8 viewing angles (Yu et al. 2020)

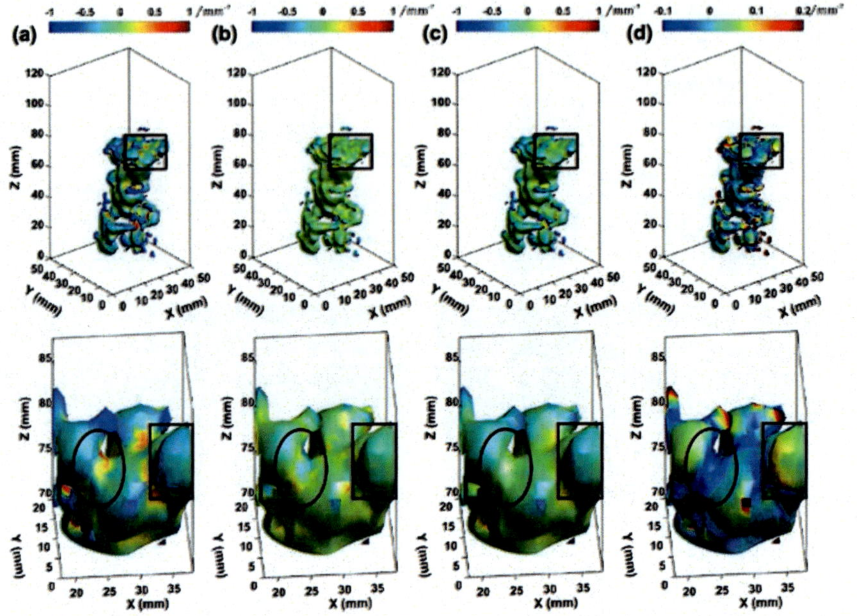

Fig. 7.33 Surface curvature results derived from traingle-based mesh approach **a** principal curvature, κ_1, **b** principal curvature, κ_2, **c** mean curvature, κ_m, and **d** Gaussian curvature, κ_G (Yu et al. 2020)

more than 6 images used. As a result, a reconstructed 3D flame structure obtained by using all 8 images of different viewing angles was used to extract the point cloud associated with the flame surface. With this surface point cloud, a triangle-based mesh approach was used to generate the surfaces that allowed the calculations of the principal curvatures, κ_1 and κ_2, as well as the mean and Gaussian curvatures, κ_m and κ_G. An example of these results can be seen in Fig. 7.33, where a selected flame surface was zoomed-in for a closer look. It is clear that these results provide significant information on the flame structures and behaviour, especially when they are time-resolved here.

From the preceding and other studies, accurate assessments of flame structures, behaviour and their quantitative nature are paramount towards more efficient combustion processes and new combustion approaches. It should also be clear that different approaches may give rise to different 3D flame structures, depending on the exact experimental technique and quantities based on, even for the same flame. Nevertheless, there are ample opportunities for a light-field camera to be used as a high-resolution 3D flame structure measurement solution, despite the significantly lower framerate. In fact, some recent studies have conducted studies into how that may be possible, such as those by Bolan et al. (2014, 2015), Nien et al. (2015), George et al. (2019), Li et al. (2019b) and Qi et al. (2019), amongst others. However, they remain at various degrees of exploration and testing, and more efforts over the coming years will be required to make light-field camera-based flame measurement technique a viable competitor to existing multi-camera-based approaches. Many of the pros and cons for and against the use of light-field cameras for flame reconstructions here are similar to those described for splash/droplet tracking. Having said that, the advantage offered by a high-resolution light-field camera for flame reconstruction could have far wider implications in terms of reconstructing flame structures with much greater accuracy. The role of turbulence in combustion process cannot be understated, especially when it comes down more efficient and clean combustions. However, it will be difficult to identify and reconstruct them well based on the setups used by Wang et al. (2013, 2015), due to their very small length-scales.

Therefore, a high-resolution light-field camera could serve as a complementary approach alongside dual high-speed camera-based stereoscopic approach, at least until the framerates of light-field cameras become sufficiently high. As for the eight-camera-based setup by Yu et al. (2020), there exists a good opportunity to improve the resolution of the flame structure by employing light-field cameras whilst potentially reducing camera costs. It has been seen in an earlier chapter that a dual light-field camera-based setup is able to produce significantly better accuracy, and such a setup could potentially lead to comparable accuracy levels in the 3D flame structures here. There is no need for fibre bundles to image four viewing perspectives upon a single camera image, which leads to significantly lower resolutions during the flame reconstructions. Additionally, camera number could also potentially be reduced to as low as two light-field cameras. The only downside is that the framerate of a typical light-field camera remains substantially lower than a high-speed 2D camera, as highlighted earlier.

3D Background Oriented Schlieren

One of the more recent developments in the area of compressible flow measurements will be Background-Oriented Schlieren (BOS). The earliest description of BOS came from the work by Dalziel et al. (2000), though they termed it as synthetic Schlieren. In the same year, Raffel et al. (2000a, b) reported upon the actual implementation of BOS in the study of helicopter rotor wake visualisations. The basic premise of a BOS approach can be appreciated in Fig. 7.34, where a flow scenario exhibiting density variations is located between a background plane surface with random dot patterns imprinted and imaging devices with appropriate optical arrangements. Note that an image of the dot pattern will be captured prior to the testing and used as a reference dot pattern. With the presence of density variations during the test however, the dot locations will be displaced visually due to light refraction and an image of the displaced dot pattern may then be captured. By comparing the latter image with the reference image through the use of image cross-correlation processing similar to that used for PIV technique, dot displacements can be obtained and further post-processed to provide density field information. Following the earlier studies and due to the relative straight forward nature of the BOS technique, adoption of BOS

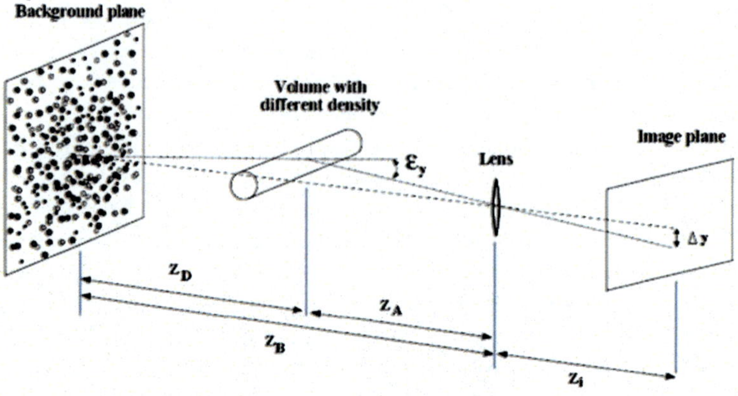

Fig. 7.34 Experimental setup associated with BOS technique (Raffel 2015)

as a more accurate 2D measurement tool for compressible flows over conventional calibrated or colour Schlieren approaches (Elsinga et al. 2004; Hargather and Settles 2012) began to garner strong interest from and take hold within the compressible flow research community. This can be seen in the works of Loose et al. (2000), Popova (2004), Raghunath et al. (2004), Jensen et al. (2005), Vasudeva et al. (2005), Ramanah and Mee (2006), Ramanah et al. (2007), Popova et al. (2008), Atcheson et al. (2009), Roosenboom and Schröder (2009, 2010), Schröder et al. (2009), Yevtikhiyeva et al. (2009), Alhaj and Seume (2010), Mizukaki (2010), Reinholtz et al. (2010), Bichal and Thurow (2010, 2013), Hargather and Settles (2011), Kirmse et al. (2011), Prasanna and Venkateshan (2011), Bencs et al. (2011), Wolf et al. (2012), Znamenskaya et al. (2012), Glazyrin et al. (2012), Gojani and Obayashi (2012), Meier and Roesgen (2013), Clem et al. (2013), Delmas et al. (2013), Gojani et al. (2013), Le Sant et al. (2014), Van Hinsberg and Rösgen (2014) and Tan et al. 2015), amongst many others.

More recently, multi-camera BOS technique as a viable 3D measurement tool (i.e. 3D-BOS) was further demonstrated by Nicolas et al. (2016) through the use of up to 12 non-coplanar cameras in their flow measurement scenarios. In particular, two different non-coplanar camera arrangements were studied as shown in Fig. 7.35, where the cameras were either distributed circumferentially and pointing towards the test scenario of interest with a slight tilt (see Fig. 7.38a), or they were arranged in a "cluster" with a 30° × 30° angular coverage of the test scenario. Results showed that the former camera arrangement is superior to the latter, even though co-planar camera arrangements will work just as well too. In particular, the results demonstrated the limitations in adopting 3D-BOS approach in test scenarios where optical access is limited. Another interesting approach in their study relates to the use of randomised dot patterns as shown in Fig. 7.36, where they found that a semi-randomised dot pattern that guarantees a large number of dots inside each interrogation window shown in Fig. 7.36b, worked better than a more randomised dot pattern shown in Fig. 7.36a, where a wider range of dot numbers could exist across all the interrogation windows and lead to less than satisfactory cross-correlation outcomes. Based on a

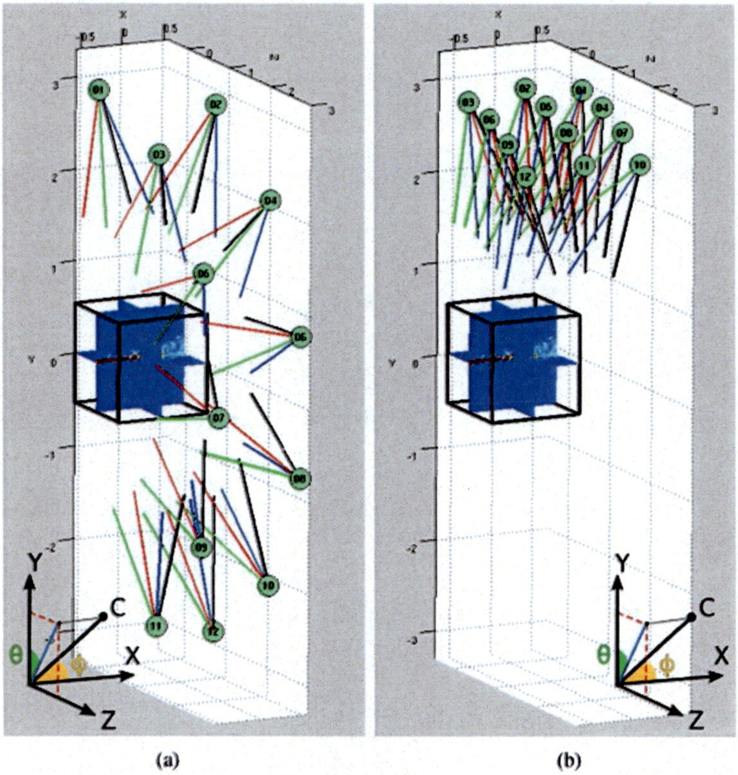

Fig. 7.35 Two different non-coplanar multi-camera arrangements for 3D-BOS technique adopted by Nicolas et al. (2016)

regularised framework for the reconstruction of 3D density volumes from the images and hence data from the multiple cameras, they were able to successfully reconstruct density fields associated with various flow fields with density variations. In particular, the investigated flow fields included a simple candle plume, heat gun hot plume, butane torch and ember, and proved the robustness and versatility of 3D-BOS across various scenarios. For instance, Fig. 7.37 shows the instantaneous isosurfaces of a hot jet emanating from the heat gun taken under optimal test conditions and it can be discerned that multi-camera-based 3D-BOS approach is highly feasible.

A follow-up study focusing on the use of similar 3D-BOS procedures but on compressible underexpanded jet flows was conducted by Nicolas et al. (2017), where instantaneous, phase-averaged and mean density fields associated with the shock structures and shear layer instabilities were determined successfully. Unlike their earlier study however, a significant number of shock structures and alternating high-low speed regions exist in these compressible jet flows, which pose additional challenges to the density measurements. Nevertheless, based on an experimental up relatively similar to their earlier study and with the use of a more powerful PIV laser

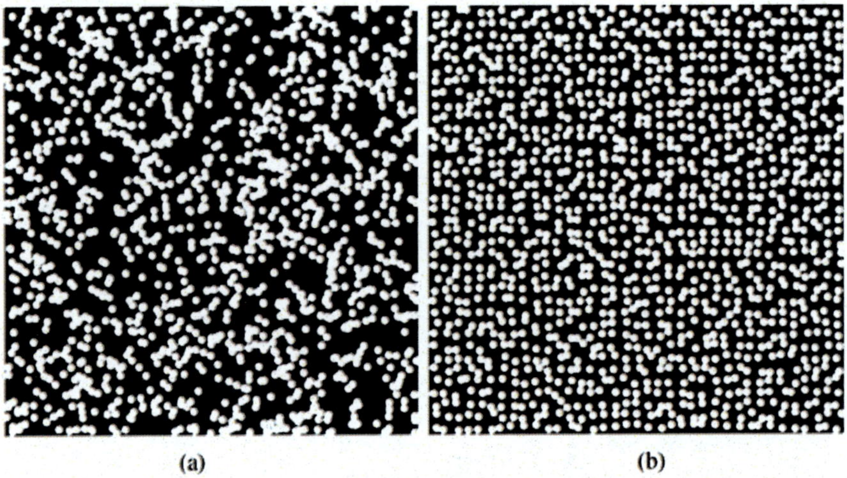

Fig. 7.36 Dot patterns generated using **a** Poisson and **b** semi-randomised distributions, the latter of which ensures a fixed number of random dots per interrogation window size (Nicolas et al. 2016)

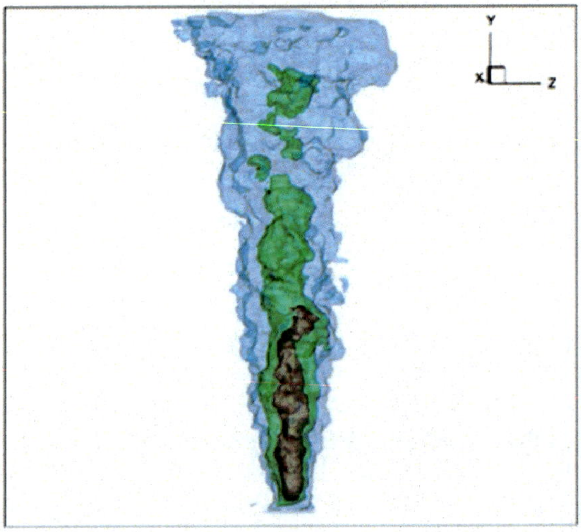

Fig. 7.37 Instantaneous density isosurfaces obtained by Nicolas et al. (2016) for a hot jet emanating from a heat gun

instead of halogen lights like in their earlier study, they were able to capture high quality 2D BOS images of the high-speed jet flows successfully. Figure 7.38 shows a direct comparison between the BOS and Schlieren images, which demonstrates that both the strong shock systems and fine turbulent shear layer structures were successfully captured. The multiple 2D-BOS images taken across a range of NPR conditions were then processed and reconstructed to arrive at the mean 3D density

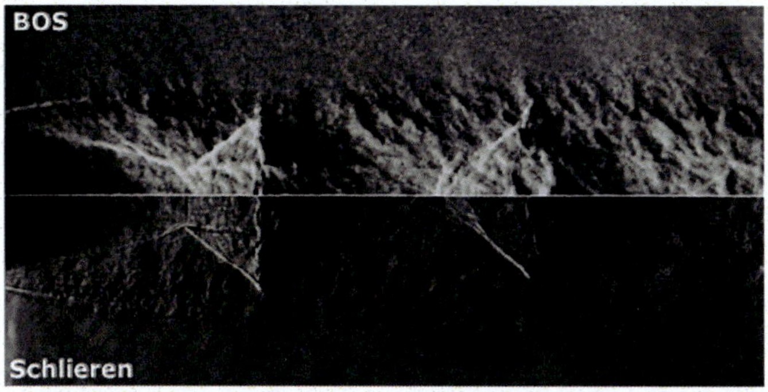

Fig. 7.38 A comparison between BOS and Schlieren images taken for a compressible jet flow at NPR = 5 (Nicolas et al. 2017)

fields as shown in Fig. 7.39. The results revealed that intricate changes to the shock cells and density variations across the expansion fans and reflect shock waves can be captured. Other than quantifying the density variations, shock cell sizes and various length-scales can be obtained from these results as well. Take for instance, comparisons between density and density gradient fields from experiments and simulations as presented in Fig. 7.40 show that there exist good agreements between the two, not only in terms of the density gradient magnitudes but in terms of quantifications of the shock cell sizes.

The tremendous usefulness of 3D-BOS approach as an accurate density field measurement tool is well established at this point. However, it should also be highlighted that the simulations conducted by Nicolas et al. (2016) demonstrated that utilising a larger number of cameras produces higher accuracy levels. Clearly, this requirement is important especially for scenarios involving highly complex/convoluted 3D test geometries or flow behaviour, where multiple cameras are to be positioned appropriately around these test geometries or flow scenarios.

This will in fact be exacerbated if the flow behaviour is highly unsteady or turbulent and high-speed cameras are required to capture the flow unsteadiness and turbulent flow structures in a time-resolved manner. However, high-speed and high-resolution cameras remain costly, not to mention the higher power requirements for the illumination sources when more cameras are used, as well as complexities in terms of setting up, positional adjustment and synchronisation of the multiple cameras.

It should not come as a surprise then, that it is at around this time that efforts to make use of a single light-field camera instead of multiple cameras for 3D-BOS measurements began in earnest. Using a single light-field camera with 3D depth sensing and thus measurement capabilities resolve many of the challenges associated with multi-camera-based 3D-BOS approach. Firstly, limited optical access may mean that a much smaller number of conventional 2D cameras can possibly be used, whilst an even smaller number of or even a single light-field cameras may be sufficient. Secondly, 3D-BOS reconstruction will be simpler for a light-field camera as compared to multiple 2D cameras, since 3D depth information can be extracted

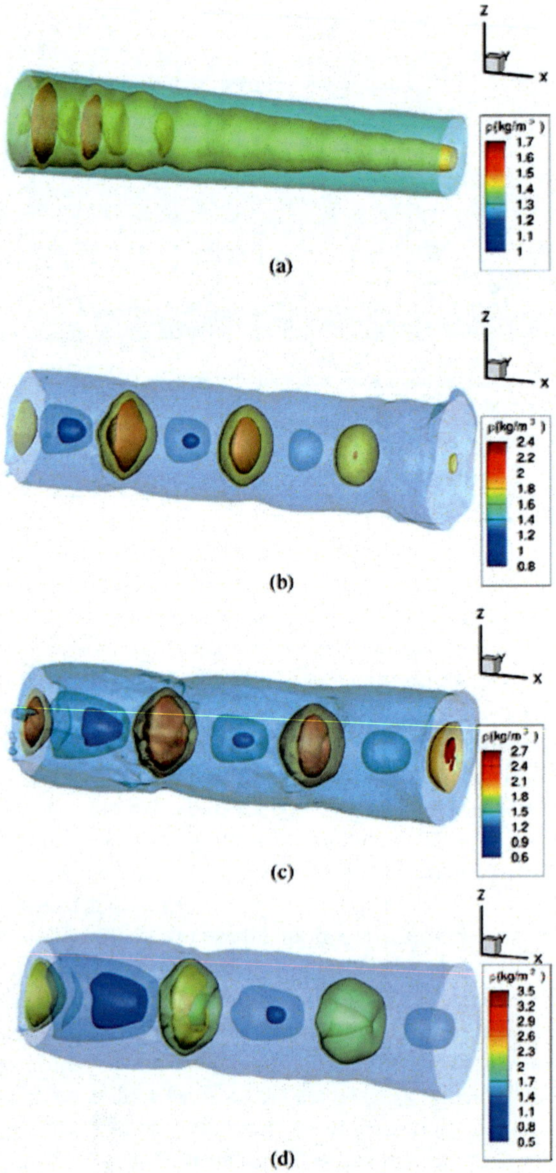

Fig. 7.39 Mean 3D density isosurfaces obtained for a compressible jet flow at **a** NPR = 2.1, **b** NPR = 3, **c** NPR = 4 and **d** NPR = 5 (Nicolas et al. 2017)

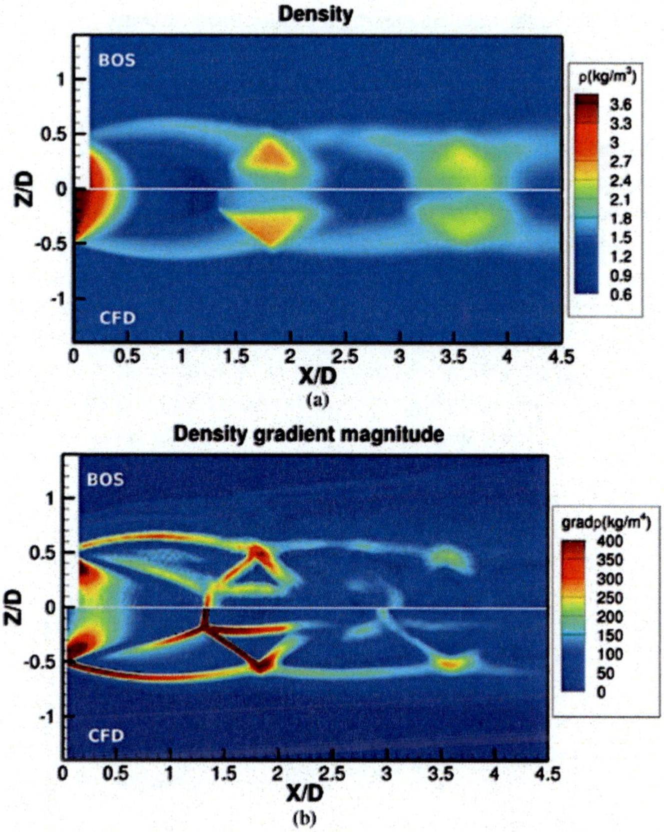

Fig. 7.40 Comparisons between density and density gradient fields obtained by BOS experiments and simulations (Nicolas et al. 2017)

directly from the light-field images for reconstruction purposes. In relation, calibration procedures will also be more straightforward for a light-field camera than multiple cameras with multiple dot pattern backgrounds from various orientations. Thirdly, data storage requirements will be significantly reduced for the use of a light-field camera instead of multiple 2D cameras. However, light-field camera is not without its disadvantages and that has to do with its significantly lower image acquisition speed and lower effective resolution. Thus, it will not be suitable for density field measurement scenarios where high-speed and/or high-resolution measurements are desired. Nevertheless, efforts have and are still being made in adapting light-field cameras for a more cost-effective and efficient 3D-BOS measurement solution.

Klemkowsky et al. (2017) proposed a "plenoptic BOS" approach whereby instead of using multiple 2D cameras or a single 2D camera measuring across different planes, a single light-field camera may accomplish the same through its unique ability to refocus along a specific plane after the image capture process. The procedures involve a light-field camera being used in-lieu of multiple 2D cameras, with the former capturing light-field images of a randomised background pattern with and without the test scenario. Note that (Klemkowsky et al. 2017) made use of a unique wavelet-based background pattern, designed such that optical distortions associated with a wider range of length/time scales of the variable density flow field can be detected by the light-field camera. Subsequently, a same number of perspective views were generated from the two light-field images captured with and without the test scenario, where image-pairs for the same perspective view are then subjected to image cross-correlation to arrive at the BOS displacement field. Thereafter, the BOS displacement fields are utilised to generate the focused BOS images, where a new refocusing function was proposed by the authors to ascertain the displacement vector fields. This new light-field camera-based BOS approach was subjected to a direct comparison with conventional BOS approach by Klemkowsky et al. (2019), when they made use of both approaches to study buoyant thermal plumes at the same time through the use of beam splitter to split the optical path from the thermal plumes to a conventional 2D and light-field camera 50–50% simultaneously. As part of the comparison shown in Fig. 7.41, they not only compared the results from the conventional BOS and light-field camera-based BOS focused along the same plane, but also compared them to down-sampled convention BOS result that matched the microlens number of the light-field camera as well.

Clearly, Fig. 7.41 shows that the resolution of the result obtained through focused BOS is significantly lower than that of the conventional BOS. However, this can be explained by the fact that the resolution achievable by the conventional BOS is a direct function of the camera sensor resolution the authors used (i.e. 6600 × 4400 pixels), whilst that of the focused BOS depends on the resolution of the microlens instead (i.e. 471 × 362 microlens). As a result, conventional BOS will produce much higher resolution results than focused BOS in the first instance. However, if one were to down-sample the result in Fig. 7.41a from its original resolution to match that of the microlens used in focused BOS here, the result presented in Fig. 7.41c actually shows a worse outcome as compared to focused BOS. Hence, despite the seemingly low microlens resolution, the depth-sensing capabilities of the light-field camera led to a measurement resolution that exceeds what the microlens resolution may suggest.

To further compare the relative accuracy levels between conventional BOS and light-field camera-based BOS approaches, displacement line profiles averaged between 0.1 and 4 mm above the plume nozzle exit but expressed in both pixels and millimetres of the image sensor by Klemkowsky et al. (2019) are shown in Fig. 7.42. The large discrepancies in the measured displacements in pixels between conventional BOS approach and light-field camera-based BOS approaches are clearly present in Fig. 7.42a, where the former registered about an order-of-magnitude larger displacement. Interestingly, the discrepancies become significantly smaller if the image sensor size was taken into account, as can be seen in Fig. 7.42b. This serves as a cautious reminder that special care and attention need to be paid towards comparing between conventional and light-field camera-based BOS accuracy levels and hence results. For more details on other comparisons between conventional and light-field camera-based BOS approaches, readers are advised to refer to Klemkowsky et al. (2019). As light-field camera-based BOS approach remains relatively novel and

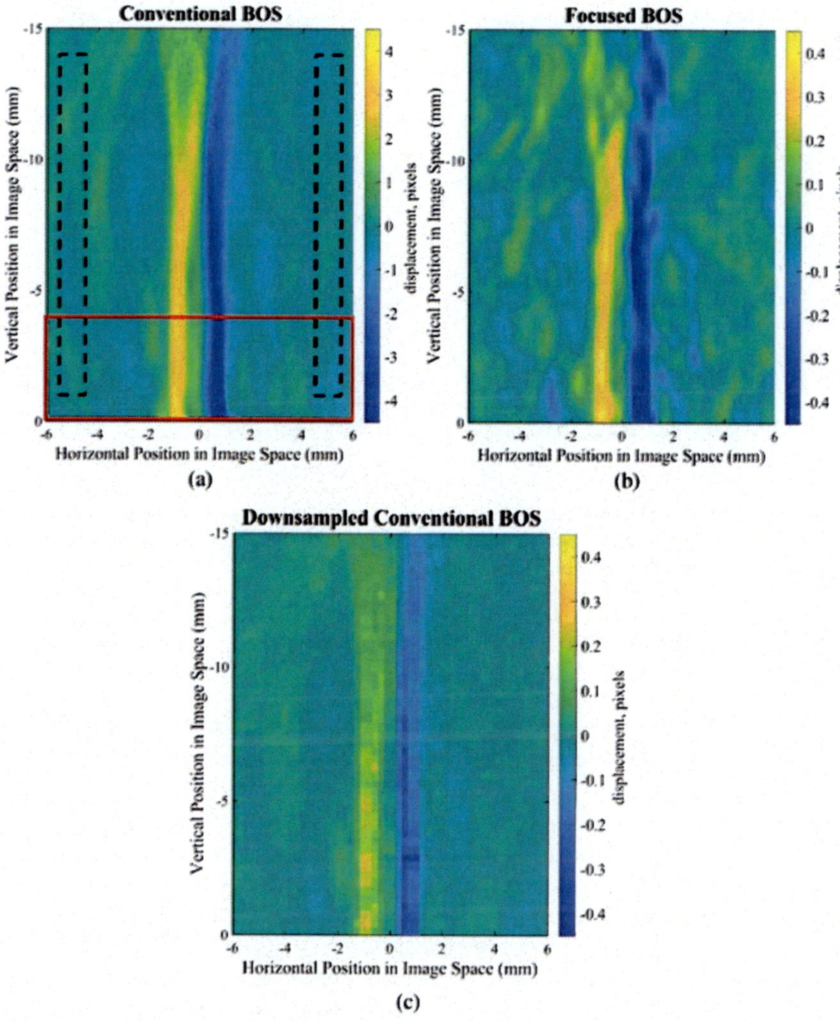

Fig. 7.41 A comparison between displacement pixels as measured by **a** conventional 2D-BOS, **b** light-field camera-based focused BOS and **c** down-sampling the result in (**a**) by matching the resolution to the number of microlens in the light-field camera used in (**b**) (Klemkowsky et al. 2019)

camera technology continues to improve, there exist significant opportunities to push the boundaries of this approach.

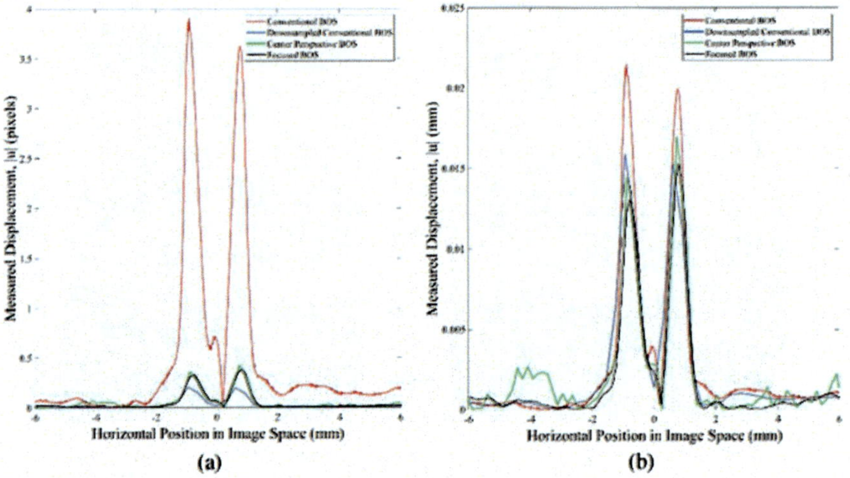

Fig. 7.42 Comparisons of averaged displacement line profiles across the plume taken between 0.1 and 4 mm above the nozzle exit for various BOS approaches, expressed in terms of **a** pixels and **b** millimetres of the image sensor (Klemkowsky et al. 2019)

Summary

Several research areas where light-field camera-based approaches have been shown to be or put forward as feasible alternatives to mult-camera-based approaches, have been briefly covered in this chapter to showcase how light-field technology may be able to push the boundaries of non-intrusive, contactless measurement techniques that are applicable towards a wide range of engineering measurement problems. For some areas such as surface temperature and metrology related measurements, light-field technologies have made significant headways that demonstrate their good performance against more conventional approaches, at potentially lower costs and lower optical access requirements. On the other hand, other measurement scenarios such as combustion processes, flame structure reconstructions and 3D-BOS, there remain significant opportunities to further understand how light-field cameras could be incorporated and how the unique depth information extracted from them could be better post-processed to improve the fidelity of the measurements.

References

Ackerman S (1962) A review of automatic radiometric pyrometry. In: Temperature; its measurement and control in science and industry, vol 2, p 839

Adhikari D, Longmire EK (2012) Visual hull method for tomographic PIV measurement of flow around moving objects. Exp Fluids 53:943–964

Alhaj O, Seume JR (2010) Optical investigation of profile losses in a linear turbine cascade

Araújo A (2017) Multi-spectral pyrometry—a review. Meas Sci Technol 28:82002
Atcheson B, Heidrich W, Ihrke I (2009) An evaluation of optical flow algorithms for background oriented schlieren imaging. Exp Fluids 46:467–476
Bencs P, Szabó S, Bordás R et al (2011) Synchronization of particle image velocimetry and background oriented schlieren measurement techniques. In: Proceedings of the 8th Pacific symposium on flow visualization and image processing, pp 1–6
Bichal A, Thurow B (2010) Development of a background oriented Schlieren-based wavefront sensor for aero-optics. In: 40th fluid dynamics conference and exhibit, p 4842
Bichal A, Thurow BS (2013) On the application of background oriented schlieren for wavefront sensing. Meas Sci Technol 25:15001
Bohn D, Krewinkel R (2009) Conjugate calculation of effusion cooling with realistic cooling hole geometries. In: Turbo expo: power for land, sea, and air, pp 31–40
Bolan J, Johnson KC, Thurow BS (2014) Preliminary investigation of three-dimensional flame measurements with a plenoptic camera. In: 30th AIAA aerodynamic measurement technology and ground testing conference, p 2520
Bolan J, Johnson KC, Thurow BS (2015) Enhanced imaging of reacting flows using 3D deconvolution and a plenoptic camera. In: 53rd AIAA aerospace sciences meeting, p 532
Cashdollar KL (1979) Three-wavelength pyrometer for measuring flame temperatures. Appl Opt 18:2595–2597
Clem M, Brown C, Fagan A (2013) Background oriented schlieren implementation in a jet-surface interaction test. In: 51st AIAA aerospace sciences meeting including the new horizons forum and aerospace exposition, p 38
Dalziel SB, Hughes GO, Sutherland BR (2000) Whole-field density measurements by 'synthetic schlieren'. Exp Fluids 28:322–335
Delmas A, Le Maoult Y, Buchlin J-M et al (2013) Shape distortions induced by convective effect on hot object in visible, near infrared and infrared bands. Exp Fluids 54:1–16
Desclaux J, Serre J (2003) M88-2E4: advanced new generation engine for Rafale multirole fighter. In: AIAA international air and space symposium and exposition: the next 100 years, p 2610
Ding J, Li H, Ma H et al (2019) A novel light field imaging based 3D geometry measurement technique for turbomachinery blades. Meas Sci Technol 30:115901
Elsinga GE, Van Oudheusden BW, Scarano F, Watt DW (2004) Assessment and application of quantitative schlieren methods: calibrated color schlieren and background oriented schlieren. Exp Fluids 36:309–325
Feng M, Gilani SZ, Wang Y, Mian A (2018) 3D face reconstruction from light field images: a model-free approach. In: Proceedings of the European conference on computer vision (ECCV), pp 501–518
Filatyev SA, Thariyan MP, Lucht RP, Gore JP (2007) Simultaneous stereo particle image velocimetry and double-pulsed planar laser-induced fluorescence of turbulent premixed flames. Combust Flame 150:201–209. https://doi.org/10.1016/j.combustflame.2007.02.005
Floyd J, Kempf AM (2011) Computed tomography of chemiluminescence (CTC): high resolution and instantaneous 3-D measurements of a matrix burner. Proc Combust Inst 33:751–758
Fu T, Cheng X, Shi C et al (2006) The set-up of a vision pyrometer. Meas Sci Technol 17:659
Fu T, Duan M, Tian J, Shi C (2016) Inverse analysis of non-uniform temperature distributions using multispectral pyrometry. Infrared Phys Technol 76:504–509
Gardner JL, Jones TP, Sainty WG (1982) Induced-transmission interference-filter array for multiwavelength pyrometry. Appl Opt 21:1259–1261
George J, Clifford C, Jenkins T, Thurow B (2019) Volumetric spectral imaging and two-color pyrometry of flames using plenoptic cameras. In: Applied optical metrology, vol III, p 1110216
Glazyrin FN, Znamenskaya IA, Mursenkova IV et al (2012) Study of shock-wave flows in the channel by schlieren and background oriented schlieren methods. Optoelectron Instrum Data Process 48:303–310
Gojani AB, Obayashi S (2012) Assessment of some experimental and image analysis factors for background-oriented schlieren measurements. Appl Opt 51:7554–7559

Gojani AB, Kamishi B, Obayashi S (2013) Measurement sensitivity and resolution for background oriented schlieren during image recording. J Vis 16:201–207

Gritsch M, Schulz A, Wittig S (2001) Effect of crossflows on the discharge coefficient of film cooling holes with varying angles of inclination and orientation. In: Turbo expo: power for land, sea, and air, p V003T01A020

Gritsch M, Colban W, Schär H, Döbbeling K (2005) Effect of hole geometry on the thermal performance of fan-shaped film cooling holes

Han J-C, Rallabandi A (2010) Turbine blade film cooling using PSP technique. Front Heat Mass Transf 1:

Hargather MJ, Settles GS (2012) A comparison of three quantitative schlieren techniques. Opt Lasers Eng 50:8–17

Hargather MJ, Settles GS (2011) Background-oriented schlieren visualization of heating and ventilation flows: HVAC-BOS. HVAC&R Res 17:771–780

Hartung G, Hult J, Balachandran R et al (2009) Flame front tracking in turbulent lean premixed flames using stereo PIV and time-sequenced planar LIF of OH. Appl Phys B 96:843–862

Heber S, Yu W, Pock T (2017) Neural epi-volume networks for shape from light field. In: Proceedings of the IEEE international conference on computer vision, pp 2252–2260

Horstmeyer R, Euliss G, Athale R, Levoy M (2009) Flexible multimodal camera using a light field architecture. In: 2009 IEEE international conference on computational photography (ICCP), pp 1–8

Hossain MM, Lu G, Sun D, Yan Y (2013) Three-dimensional reconstruction of flame temperature and emissivity distribution using optical tomographic and two-colour pyrometric techniques. Meas Sci Technol 24:74010. https://doi.org/10.1088/0957-0233/24/7/074010

Huang J, Liu H, Wang Q, Cai W (2020) Limited-projection volumetric tomography for time-resolved turbulent combustion diagnostics via deep learning. Aerosp Sci Technol 106:106123. https://doi.org/10.1016/j.ast.2020.106123

Im S, Jeon YJ, Sung HJ (2015) Tomo-PIV measurement of flow around an arbitrarily moving body with surface reconstruction. Exp Fluids 56:1–16

Jensen OS, Kunsch JP, Rösgen T (2005) Optical density and velocity measurements in cryogenic gas flows. Exp Fluids 39:48–55

Jeon YJ, Sung HJ (2012) Three-dimensional PIV measurement of flow around an arbitrarily moving body. Exp Fluids 53:1057–1071

Johannsen O, Sulc A, Goldluecke B (2016) What sparse light field coding reveals about scene structure. In: Proceedings of the IEEE conference on computer vision and pattern recognition, pp 3262–3270

Kelly D, Scarborough D, Thurow B (2021) A novel multi-band plenoptic pyrometer for high temperature applications. Meas Sci Technol

Kirmse T, Agocs J, Schröder A et al (2011) Application of particle image velocimetry and the background-oriented schlieren technique in the high-enthalpy shock tunnel Göttingen. Shock Waves 21:233–241

Klemkowsky JN, Fahringer TW, Clifford CJ et al (2017) Plenoptic background oriented schlieren imaging. Meas Sci Technol 28:95404

Klemkowsky JN, Clifford CJ, Bathel BF, Thurow BS (2019) A direct comparison between conventional and plenoptic background oriented schlieren imaging. Meas Sci Technol 30:64001

Le Sant Y, Todoroff V, Bernard-Brunel A et al (2014) Multi-camera calibration for 3DBOS. In: 17th international symposium on applications of laser techniques to fluid mechanics

Li D, Feng C, Daniel K et al (2019a) Turbine blade temperature error as measured with an optical pyrometer under different wavelengths and blade TBC thickness. Appl Opt 58:1626–1630

Li J, Hossain MM, Sun J et al (2019b) Simultaneous measurement of flame temperature and absorption coefficient through LMBC-NNLS and plenoptic imaging techniques. Appl Therm Eng 154:711–725

Liu H, Zheng S (2020) Measurement of temperature and wavelength-dependent emissivity distributions using multi-wavelength radiation thermometry. Opt Commun 472:125895

Loose S, Richard H, Dewhirst T, Raffel M (2000) Background oriented schlieren (BOS) and particle image velocimetry (PIV) applied for transonic turbine blade investigations. In: 10th international symposium on applications of laser techniques to fluid mechanics. Lisbon, Portugal, p 24

Lu G, Yan Y (2006) Temperature profiling of pulverized coal flames using multicolor pyrometric and digital imaging techniques. IEEE Trans Instrum Meas 55:1303–1308

Lü Y, He X, Wei Z-H et al (2016) Ambient temperature-independent dual-band mid-infrared radiation thermometry. Appl Opt 55:2169–2174

Luan Y, Mei D, Shi S (2021) Light-field multi-spectral radiation thermometry. Opt Lett 46:9–12

Ma H, Qian Z, Mu T, Shi S (2019) Fast and accurate 3D measurement based on light-field camera and deep learning. Sensors 19:4399

Meier AH, Roesgen T (2013) Improved background oriented schlieren imaging using laser speckle illumination. Exp Fluids 54:1–6

Mendelson L, Techet AH (2015) Quantitative wake analysis of a freely swimming fish using 3D synthetic aperture PIV. Exp Fluids 56:1–19

Meng X, Diebold GJ (2016) Optical pyrometer based on the gas phase photoacoustic effect. Opt Lett 41:2221–2224

Mizukaki T (2010) Visualization of compressible vortex rings using the background-oriented schlieren method. Shock Waves 20:531–537

Moeck JP, Bourgouin J-F, Durox D et al (2013) Tomographic reconstruction of heat release rate perturbations induced by helical modes in turbulent swirl flames. Exp Fluids 54:1–17

Montomoli F, Massini M, Salvadori S, Martelli F (2012) Geometrical uncertainty and film cooling: fillet radii

Ng WB, Zhang Y (2003) Stereoscopic imaging and reconstruction of the 3D geometry of flame surfaces. Exp Fluids 34:484–493

Nicolas F, Todoroff V, Plyer A et al (2016) A direct approach for instantaneous 3D density field reconstruction from background-oriented schlieren (BOS) measurements. Exp Fluids 57:13

Nicolas F, Donjat D, Léon O et al (2017) 3D reconstruction of a compressible flow by synchronized multi-camera BOS. Exp Fluids 58:46

Nien H, Fessler JA, Sick V (2015) Model-based image reconstruction of chemiluminescence using a plenoptic 2.0 camera. In: 2015 IEEE international conference on image processing (ICIP). pp 359–363

Petersson P, Olofsson J, Brackman C et al (2007) Simultaneous PIV/OH-PLIF, Rayleigh thermometry/OH-PLIF and stereo PIV measurements in a low-swirl flame. Appl Opt 46:3928–3936. https://doi.org/10.1364/AO.46.003928

Popova EM (2004) Processing schlieren-background patterns by constructing the direction field. J Opt Technol 71:572–574

Popova E, Kompenhans J, Skornyakova N (2008) Investigation of the accuracy of the background oriented schlieren method

Prasanna S, Venkateshan SP (2011) Construction of two dimensional temperature field from first derivative fields. Exp Therm Fluid Sci 35:1019–1029

Qi Q, Hossain MM, Zhang B et al (2019) Flame temperature reconstruction through a multi-plenoptic camera technique. Meas Sci Technol 30:124002

Raffel M (2015) Background-oriented schlieren (BOS) techniques. Exp Fluids 56:1–17

Raffel M, Richard H, Meier GEA (2000a) On the applicability of background oriented optical tomography for large scale aerodynamic investigations. Exp Fluids 28:477–481

Raffel M, Tung C, Richard H et al (2000b) Background oriented stereoscopic schlieren (BOSS) for full scale helicopter vortex characterization. In: 9th international symposium on flow visualization, pp 23–24

Raghunath S, Mee DJ, Roesgen T, Jacobs PA (2004) Visualization of supersonic flows in shock tunnels, using the background oriented schlieren technique

Ramanah D, Mee D (2006) Scramjet flow visualization using background oriented schlieren in hypersonic impulse facilities. In: 14th AIAA/AHI space planes and hypersonic systems and technologies conference, p 8004

Ramanah D, Raghunath S, Mee DJ et al (2007) Background oriented schlieren for flow visualisation in hypersonic impulse facilities. Shock Waves 17:65–70

Reinholtz C, Heltsley F, Scott K, Rhode M (2010) Visualization of jettison motor plumes from an Orion launch abort vehicle wind tunnel model using background-oriented schlieren. In: US air force T&E days 2010, p 1736

Roosenboom EWM, Schröder A (2009) Qualitative investigation of a propeller slipstream with background oriented Schlieren. J vis 12:165–172

Roosenboom E, Schröder A (2010) Image-based measurement techniques of increased complexity for industrial propeller flow investigations. In: 27th AIAA aerodynamic measurement technology and ground testing conference, p 4210

Schröder A, Over B, Geisler R et al (2009) Measurements of density fields in micro nozzle plumes in vacuum by using an enhanced tomographic background oriented schlieren (BOS) technique. In: 9th international symposium on measurement technology and intelligent instruments, Saint-Petersburg

Sheng H, Zhao P, Zhang S et al (2018) Occlusion-aware depth estimation for light field using multi-orientation EPIs. Pattern Recognit 74:587–599

Shin C, Jeon H-G, Yoon Y et al (2018) Epinet: a fully-convolutional neural network using epipolar geometry for depth from light field images. In: Proceedings of the IEEE conference on computer vision and pattern recognition, pp 4748–4757

Stratton BJ (2005) Determining flame height and flame pulsation frequency and estimating heat release rate from 3D flame reconstruction

Su L, Zhou Z, Yuan Y et al (2015) A snapshot light field imaging spectrometer. Optik (stuttg) 126:877–881

Sun XG, Yuan GB, Dai JM, Chu ZX (2005) Processing method of multi-wavelength pyrometer data for continuous temperature measurements. Int J Thermophys 26:1255–1261

Tan DJ, Edgington-Mitchell D, Honnery D (2015) Measurement of density in axisymmetric jets using a novel background-oriented schlieren (BOS) technique. Exp Fluids 56:1–11

Tsyba GA, Salamatov VG, Polyakov VL (2003) A video pyrometer. Instrum Exp Tech 46:480–483

Van Hinsberg NP, Rösgen T (2014) Density measurements using near-field background-oriented schlieren. Exp Fluids 55:1–11

Vasudeva G, Honnery DR, Soria J (2005) Non-intrusive measurement of a density field using the background oriented schlieren (BOS) method. In: Proceedings of the Australian conference on laser diagnostic in fluid mechanics and combustion

Wang Q, Zhang Y (2011) High speed stereoscopic shadowgraph imaging and its digital 3D reconstruction. Meas Sci Technol 22:65302. https://doi.org/10.1088/0957-0233/22/6/065302

Wang Q, Huang HW, Zhang Y, Zhao C (2013) Impinging flame ignition and propagation visualisation using Schlieren and colour-enhanced stereo imaging techniques. Fuel 108:177–183. https://doi.org/10.1016/j.fuel.2013.01.048

Wang Q, Yang J, Huang HW et al (2014) Three-dimensional investigation of the dynamics of a propane diffusion flame. Fuel 116:448–454. https://doi.org/10.1016/j.fuel.2013.08.038

Wang Q, Zhao CY, Zhang Y (2015) Time-resolved 3D investigation of the ignition process of a methane diffusion impinging flame. Exp Therm Fluid Sci 62:78–84

Wang Q, Yu T, Liu H et al (2020) Optimization of camera arrangement for volumetric tomography with constrained optical access. J Opt Soc Am B 37:1231–1239. https://doi.org/10.1364/JOSAB.385291

Weinkauff J, Trunk P, Frank JH et al (2015) Investigation of flame propagation in a partially premixed jet by high-speed-Stereo-PIV and acetone-PLIF. Proc Combust Inst 35:3773–3781. https://doi.org/10.1016/j.proci.2014.05.022

Wen C-D, Mudawar I (2004) Emissivity characteristics of roughened aluminum alloy surfaces and assessment of multispectral radiation thermometry (MRT) emissivity models. Int J Heat Mass Transf 47:3591–3605

Wolf C, Klei C, Buffo R et al (2012) Comparison of rocket near-wakes with and without nozzle simulation in subsonic freestream conditions. In: 42nd AIAA fluid dynamics conference and exhibit, p 3019

Wu Y, Wang Q, Zhao CY (2020) Three-Dimensional droplet splashing dynamics measurement with a stereoscopic shadowgraph system. Int J Heat Fluid Flow 83:108576. https://doi.org/10.1016/j.ijheatfluidflow.2020.108576

Xing J, Peng B, Ma Z et al (2017) Directly data processing algorithm for multi-wavelength pyrometer (MWP). Opt Express 25:30560–30574

Yan W, Panahi A, Levendis YA (2020) Spectral emissivity and temperature of heated surfaces based on spectrometry and digital thermal imaging–validation with thermocouple temperature measurements. Exp Therm Fluid Sci 112:110017

Yevtikhiyeva OA, Skornyakova NM, Udalov AV (2009) An investigation of the error of the background schlieren method. Meas Tech 52:1300–1305

Yu T, Ruan C, Liu H et al (2018) Time-resolved measurements of a swirl flame at 4 kHz via computed tomography of chemiluminescence. Appl Opt 57:5962–5969

Yu T, Wang Q, Ruan C et al (2020) A quantitative evaluation method of 3D flame curvature from reconstructed flame structure. Exp Fluids 61:1–13

Zhang Z (2000) A flexible new technique for camera calibration. IEEE Trans Pattern Anal Mach Intell 22:1330–1334. https://doi.org/10.1109/34.888718

Zhao Y, Li H, Mei D, Shi S (2020) Metric calibration of unfocused plenoptic cameras for three-dimensional shape measurement. Opt Eng. https://doi.org/10.1117/1.oe.59.7.073104

Znamenskaya I, Vinnichenko N, Glazyrin F (2012) Quantitative measurements of the density gradient on the flat shock wave by means of background oriented schlieren. In: Proceedings of the 15th international symposium on flow visualisation, Minsk

List of Formulas

$$I = I(x, y, z, \theta, \varnothing, \lambda, t) \quad (1.1)$$

$$I = I(x, y, z, \theta, \varnothing) \quad (1.2)$$

$$I = I(u, v, s, t,) \quad (1.3)$$

$$\frac{1}{z_o} + \frac{1}{z_i} = \frac{1}{f_m} \quad (2.1)$$

$$M = -\frac{z_i}{z_o} = -\frac{x_i}{x_o} = -\frac{y_i}{y_o} \quad (2.2)$$

$$\frac{1}{S_o} + \frac{1}{S_i} = \frac{1}{f_m} \quad (2.3)$$

$$M = -\frac{S_i}{S_o} \quad (2.4)$$

$$s = -Mx_o \quad (2.5)$$

$$t = -Mx_o \quad (2.6)$$

$$M_{\text{chief\_ray}} = -\frac{S_i}{z_o} \quad (2.7)$$

$$\frac{1}{z_o} + \frac{1}{S_i} = \frac{1}{f_m} \quad (2.8)$$

$$\text{DOF} = z_{o,\text{far}} - z_{o,\text{near}} \quad (2.9)$$

S. Shi and T. H. New (eds.), *Development and Application of Light-Field Cameras in Fluid Measurements*, https://doi.org/10.1007/978-3-031-01779-7

$$z_{o,\text{far}} = \frac{z_o p_m}{p_m - D_{CoC}} \tag{2.10}$$

$$z_{o,\text{near}} = \frac{Z_o p_m}{p_m + D_{CoC}} \tag{2.11}$$

$$I(s,t) = \iint\limits_{p_m} L(u,v,s,t)\mathrm{d}u\,\mathrm{d}v \tag{2.12}$$

$$\frac{p_m}{S_i} = \frac{p_l}{d_{i,l}} = \frac{p_l}{f_l} \tag{2.13}$$

$$(f/\#)_m = (f/\#)_l/(1-M) \tag{2.14}$$

$$I_{u_0,v_0}(s,t) = L(s,t,u=u_0,v=v_0) \tag{2.15}$$

$$I\left(s',t'\right) = \iint\limits_{p_m} L\left(u\left(1-\frac{1}{\alpha}\right)+\frac{s'}{\alpha}, v\left(1-\frac{1}{\alpha}\right)+\frac{t'}{\alpha}, u, v\right)\mathrm{d}u\,\mathrm{d}v \tag{2.16}$$

$$\begin{pmatrix} x' \\ y' \\ \theta' \\ \varphi' \end{pmatrix} = \begin{pmatrix} 1 & 0 & z_o & 0 \\ 0 & 1 & 0 & z_o \\ 0 & 0 & 1 & 0 \\ 0 & 0 & 0 & 1 \end{pmatrix} \begin{pmatrix} x \\ y \\ \theta \\ \varphi \end{pmatrix} \tag{2.17}$$

$$\begin{pmatrix} x' \\ y' \\ \theta' \\ \varphi' \end{pmatrix} = \begin{pmatrix} 1 & 0 & 0 & 0 \\ 0 & 1 & 0 & 0 \\ -1/f_m & 0 & 1 & 0 \\ 0 & -1/f_m & 0 & 1 \end{pmatrix} \begin{pmatrix} x \\ y \\ \theta \\ \varphi \end{pmatrix} \tag{2.18}$$

$$\begin{pmatrix} x' \\ y' \\ \theta' \\ \varphi' \end{pmatrix} = \begin{pmatrix} 1 & 0 & S_i & 0 \\ 0 & 1 & 0 & S_i \\ 0 & 0 & 1 & 0 \\ 0 & 0 & 0 & 1 \end{pmatrix} \begin{pmatrix} x \\ y \\ \theta \\ \varphi \end{pmatrix} \tag{2.19}$$

$$\begin{pmatrix} x' \\ y' \\ \theta' \\ \varphi' \end{pmatrix} = \begin{pmatrix} 1 & 0 & 0 & 0 \\ 0 & 1 & 0 & 0 \\ -1/f_l & 0 & 1 & 0 \\ 0 & -1/f_l & 0 & 1 \end{pmatrix} \begin{pmatrix} x \\ y \\ \theta \\ \varphi \end{pmatrix} + \begin{pmatrix} 0 \\ 0 \\ S_x/f_l \\ S_y/f_l \end{pmatrix} \tag{2.20}$$

$$\begin{pmatrix} x' \\ y' \\ \theta' \\ \varphi' \end{pmatrix} = \begin{pmatrix} 1 & 0 & f_l & 0 \\ 0 & 1 & 0 & f_l \\ 0 & 0 & 1 & 0 \\ 0 & 0 & 0 & 1 \end{pmatrix} \begin{pmatrix} x \\ y \\ \theta \\ \varphi \end{pmatrix} \tag{2.21}$$

$$E\left(X_j, Y_j, Z_j\right)^{k+1} = E\left(X_j, Y_j, Z_j\right)^k \left(\frac{I(x_i, y_i)}{\sum_{j \in N_i} w_{i,j} E\left(X_j, Y_j, Z_j\right)^k} \right)^{\mu w_{i,j}} \tag{3.1}$$

$$D_{CoC} = p_m \left(\frac{S_i}{S_o} - \frac{S_i}{z + S_o} \right) = p_m \left(-M - \frac{1}{\frac{z}{S_i} - \frac{1}{M}} \right) \tag{3.2}$$

$$D_{df} = \frac{S_i + f_l}{S_i} \cdot p_m \left(-M - \frac{1}{\frac{z}{S_i} - \frac{1}{M}} \right) \tag{3.3}$$

$$S_i = f_m(1 - M) \tag{3.4}$$

$$f_{\#} = f_m / p_m \tag{3.5}$$

$$D_{df} = k_1 \cdot p_m \left(-M - \frac{1}{\frac{k_2 z + k_3 x + k_4 y + k_5}{S_i} - \frac{1}{M}} \right) \tag{3.6}$$

$$y_p = S_y + \frac{y_l + S_y - y_m}{S_i} f_l = C_{l(i)} - \frac{f_l}{S_i} y_m + \frac{y_l}{S_i} f_l \tag{3.7}$$

$$p_c = C_{l(i)} - \frac{f_l}{S_i} y_c \tag{3.8}$$

$$C_{df} = \frac{y_c}{p_m} D_{df} + C_{l(i)} \tag{3.9}$$

$$\frac{p_m}{S_i} = \frac{p_{l\_\text{sensor}}}{f_l} \tag{3.10}$$

$$\left(C_{l(i)} - p_c\right) D_{df} + p_{l\_\text{sensor}} \left(C_{l(i)} - C_{df}\right) = 0 \tag{3.11}$$

$$\begin{pmatrix} C_{df}(x) \\ C_{df}(y) \\ 1 \end{pmatrix} = \mathcal{M}_{4 \times 3} \begin{pmatrix} X \\ Y \\ Z \\ 1 \end{pmatrix} \tag{3.12}$$

$$\frac{y_m}{p_m} = \frac{C_{df} - \left(C_{l(i)} + y_l \frac{S_i + f_l}{S_i}\right)}{D_{df}} \tag{3.13}$$

$$y_p = C_{l(i)} + \frac{p_{l\_\text{sensor}} \left(y_l \frac{S_i + f_l}{S_i} + C_{li} - C_{df} \right)}{D_{df}} \tag{3.14}$$

$$y_{p\_dn} = \begin{cases} C_{l(i)} + \frac{p_{l\_sensor}\left(-\frac{d_l}{2}+C_{l(i)}-C_{df}\right)}{D_{df}}, & z < 0 \\ C_{l(i)} + \frac{p_{l\_sensor}\left(\frac{d_l}{2}+C_{l(i)}-C_{df}\right)}{D_{df}}, & z > 0 \end{cases} \tag{3.15}$$

$$y_{p\_up} = \begin{cases} C_{l(i)} + \frac{p_{l\_sensor}\left(\frac{d_l}{2}\cdot+C_{l(i)}-C_{df}\right)}{D_{df}}, & z < 0 \\ C_{l(i)} + \frac{p_{l\_sensor}\left(-\frac{d_l}{2}\cdot+C_{l(i)}-C_{df}\right)}{D_{df}}, & z > 0 \end{cases} \tag{3.16}$$

$$\frac{1}{S'_o} + \frac{1}{S'_i} = \frac{1}{P^z} + \frac{1}{-Q^z} = \frac{1}{f_m} \tag{3.17}$$

$$\begin{bmatrix} Q^x \\ Q^y \\ Q^z \end{bmatrix} = \frac{-S'_i}{S'_o} \begin{bmatrix} P^x \\ P^y \\ P^z \end{bmatrix} = \frac{f_m}{f_m - P^z} \begin{bmatrix} P^x \\ P^y \\ P^z \end{bmatrix} \tag{3.18}$$

$$\begin{bmatrix} Q^u \\ Q^v \end{bmatrix} = \begin{bmatrix} O_p^u + \frac{1}{p_p} Q^x \\ O_p^v - \frac{1}{p_p}[(S_i + Q^z)\sin\gamma - Q^y \cos\gamma] \end{bmatrix} \tag{3.19}$$

$$\begin{aligned} Q^d &= \frac{1}{p_p}[Q^y \sin\gamma + (S_i + Q^z)\cos\gamma] \\ &= \frac{1}{p_p}\left[\frac{f_m}{f_m - P^z} P^y \sin\gamma + \left(S_i + \frac{f_m}{f_m - P^z} P^z\right)\cos\gamma\right] \end{aligned} \tag{3.20}$$

$$\begin{bmatrix} p_c^u \\ p_c^v \end{bmatrix} = \begin{bmatrix} h_c^u \\ h_c^v \end{bmatrix} + \frac{f_l}{Q_d}\left(\begin{bmatrix} h_c^u \\ h_c^v \end{bmatrix} - \begin{bmatrix} Q^u \\ Q^v \end{bmatrix}\right) \tag{3.21}$$

$$\begin{pmatrix} p^{up,u} \\ p^{up,v} \\ \mu'_{up} \\ \sigma'_{up} \end{pmatrix} = \begin{pmatrix} 1 & & f_l & \\ & 1 & & f_l \\ & & 1 & \\ & & & 1 \end{pmatrix} \begin{pmatrix} 1 & & & \\ & 1 & & \\ -1/f_l & & 1 & \\ & -1/f_l & & 1 \end{pmatrix} \begin{pmatrix} h_c^u \\ h_c^v + \frac{p_l}{2p_p} \\ \mu_{up} \\ \sigma_{up} \end{pmatrix} + \begin{pmatrix} h_c^u \\ h_c^v \\ 0 \\ 0 \end{pmatrix} \tag{3.22}$$

$$\begin{pmatrix} \mu_{up} \\ \sigma_{up} \end{pmatrix} = \begin{pmatrix} \frac{h_c^u - Q^u}{Q^d} \\ \frac{h_c^v + \frac{p_l}{2p_p} - Q^v}{Q^d} \end{pmatrix} \tag{3.23}$$

$$\begin{cases} p^{up} = \begin{pmatrix} h_c^u \\ h_c^v \end{pmatrix} + f_l \begin{pmatrix} \frac{h_c^u - Q^u}{Q^d} \\ \frac{h_c^v + \frac{p_l}{2p_p} - Q^v}{Q^d} \end{pmatrix} \\ p^{dn} = \begin{pmatrix} h_{ci}^u \\ h_{ci}^v \end{pmatrix} + f_l \begin{pmatrix} \frac{h_{ci}^u - Q^u}{Q^d} \\ \frac{h_{ci}^v - \frac{p_l}{2p_p} - Q^v}{Q^d} \end{pmatrix} \end{cases} \tag{3.24}$$

$$\begin{cases} \frac{QQ_1}{MM_1} = \frac{EoC_M Q_1}{EoC_M M_1} \\ \frac{QQ_1}{NN_1} = \frac{EoC_N Q_1}{EoC_N N_1} \end{cases} \tag{3.25}$$

$$C_{EoC} = \frac{1}{2}(EoC_M + EoC_N) \tag{3.26}$$

$$D_{EoC}^{\text{major}} = \frac{1}{2}|EoC_M EoC_N| \tag{3.27}$$

$$D_{EoC}^{\text{minor}} = \frac{1}{2}|EoC_G EoC_H| \tag{3.28}$$

$$\frac{g_c h_c}{l_c A_1} = \frac{f_l}{AA_1} \tag{3.29}$$

$$\frac{A_m}{\frac{A^d}{\cos\gamma}} = \frac{p_l}{f_l} = \frac{1}{f_{\#}(1-M)} \tag{3.30}$$

$$A_m = \frac{A^d}{\cos\gamma}\frac{p_l}{f_l} \tag{3.31}$$

$$\begin{cases} D_{EoC}^{\text{major}} = k \times \frac{1}{2}|EoC_M EoC_N| \\ D_{EoC}^{\text{minor}} = k \times \frac{1}{2}|EoC_G EoC_H| \end{cases} \tag{3.32}$$

$$Q^{u,v} = R^{u,v} - \frac{Q^d}{A^d}R^{u,v} \tag{3.33}$$

$$\begin{aligned} Q^d = {} & a_0 + a_1 R^u + a_2 R^v + a_3 P^z + a_4 R^u R^v + a_5 R^u P^z + a_6 \left(P^z\right)^2 + a_7 R^v P^z \\ & + a_8 (R^v)^2 + a_9 \left(R^u\right)^2 + a_{10}\left(R^u\right)^2 R^v + a_{11}(R^v)^2 R^u + a_{12}\left(R^u\right)^2 P^z \\ & + a_{13}(R^v)^2 P^z + a_{14}\left(P^z\right)^2 R^v + a_{15}\left(P^z\right)^2 R^u + a_{16} R^u R^v P^z + a_{17}\left(R^u\right)^3 \\ & + a_{18}(R^v)^3 + a_{19}\left(P^z\right)^3 \end{aligned} \tag{3.34}$$

$$\begin{pmatrix} R^u \\ R^v \\ 1 \end{pmatrix} = M_{4\times 3}\begin{pmatrix} P^x \\ P^y \\ P^z \\ 1 \end{pmatrix} \tag{3.35}$$

$$E\left(X_j, Y_j, Z_j\right)^{k+1} = E\left(X_j, Y_j, Z_j\right)^k \left(\frac{I(x_i, y_i)}{\sum_{j\in N_i} w_{i,j} E\left(X_j, Y_j, Z_j\right)^k}\right)^{\mu w_{i,j}} \tag{4.1}$$

$$Q_{\text{Recon}} = \frac{\sum E_1(x,y,z)E_0(x,y,z)}{\sqrt{\sum E_1^2(x,y,z) \times E_0^2(x,y,z)}} \tag{4.2}$$

$$\begin{cases} u(x,y,z) = 25\mathrm{e}^{i\left(k_x x + k_y y + k_z z\right)} \\ v(x,y,z) = 0 \\ w(x,y,z) = 0 \end{cases} \tag{4.3}$$

$$\begin{cases} u(x, y, z) = 0 \\ v(x, y, z) = 0 \\ w(x, y, z) = 25e^{i\left(k_x x + k_y y + k_z z\right)} \end{cases} \tag{4.4}$$

$$\text{LTPR} = \frac{\left(p_x \times p_y\right)_{\text{LF-PIV}}}{\left(p_x \times p_y \times N_c\right)_{\text{Tomo-PIV}}} \tag{4.5}$$

$$\text{Error}_{\text{divergence}} = \left| \frac{\partial u}{\partial x} + \frac{\partial v}{\partial y} + \frac{\partial w}{\partial z} \right| \tag{4.6}$$

$$\frac{I_{\text{ref}}}{I} = A(T) + B(T)\frac{P}{P_{\text{ref}}} \tag{5.1}$$

$$\beta = \left(\frac{\delta_1}{\tau_w}\right)\frac{\mathrm{d}P}{\mathrm{d}x} \tag{6.1}$$

$$\frac{\partial u}{\partial x} + \frac{\partial v}{\partial y} + \frac{\partial w}{\partial z} = 0 \tag{6.2}$$

$$u(t) = u_{\max} \sin\left(\frac{2\pi t}{\tau} + \theta\right) \tag{6.3}$$

$$\overline{u} = \frac{1}{\tau} \int_0^{\tau/2} u(t)\mathrm{d}t = \frac{1}{\tau} \int_0^{\tau/2} u_{\max} \sin\left(\frac{2\pi t}{\tau}\right)\mathrm{d}t = \frac{u_{\max}}{\pi} \tag{6.4}$$

$$U_0 = \frac{\overline{u}\pi}{\sqrt{2}} \tag{6.5}$$

$$\text{AR} = \frac{l}{d} \tag{6.6}$$

$$Sr = \frac{fd}{\overline{u}} = \frac{d}{L_o} \tag{6.7}$$

$$\text{Re} = \frac{\overline{u}d}{\nu} \tag{6.8}$$

$$L_o = \frac{\overline{u}}{f} \tag{6.9}$$

$$f = 1/\tau \tag{6.10}$$

$$D \equiv \left[d_{i,j}\right] = \left[\, v_r \;\; v_{cr} \;\; v_{ci} \,\right] \begin{bmatrix} \lambda_r & & \\ & \lambda_{cr} & \lambda_{ci} \\ & -\lambda_{ci} & \lambda_{cr} \end{bmatrix} \left[\, v_r \;\; v_{cr} \;\; v_{ci} \,\right]^{-1} \tag{6.11}$$

$$I_i = A_i \varepsilon(\lambda_i, T) \frac{1}{\lambda_i^5 \left(e^{\frac{C_2}{\lambda_i T}} - 1 \right)} \quad (i = 1, 2, \ldots, 9) \tag{7.1}$$

$$\sum_{i=1}^{n} [T_i - E(T_i)]^2 = 0 \tag{7.2}$$